Aspects of
Multivariate Statistical Analysis
in Geology

Aspects of Multivariate Statistical Analysis in Geology

Richard A. Reyment and Enrico Savazzi

1999

ELSEVIER

Amsterdam – Lausanne – New York – Oxford – Shannon – Singapore – Tokyo

ELSEVIER SCIENCE B.V.
Sara Burgerhartstraat 25
P.O. Box 211, 1000 AE Amsterdam, The Netherlands

© 1999 Elsevier Science B.V. All rights reserved.

This work is protected under copyright by Elsevier Science, and the following terms and conditions apply to its use:

Photocopying
Single photocopies of single chapters may be made for personal use as allowed by national copyright laws. Permission of the Publisher and payment of a fee is required for all other photocopying, including multiple or systematic copying, copying for advertising or promotional purposes, resale, and all forms of document delivery. Special rates are available for educational institutions that wish to make photocopies for non-profit educational classroom use.

Permissions may be sought directly from Elsevier Science Rights & Permissions Department, PO Box 800, Oxford OX5 1DX, UK; phone: (+44) 1865 843830, fax: (+44) 1865 853333, e-mail: permissions@elsevier.co.uk. You may also contact Rights & Permissions directly through Elsevier's home page (http://www.elsevier.nl), selecting first 'Customer Support', then 'General Information', then 'Permissions Query Form'.

In the USA, users may clear permissions and make payments through the Copyright Clearance Center, Inc., 222 Rosewood Drive, Danvers, MA 01923, USA; phone: (978) 7508400, fax: (978) 7504744, and in the UK through the Copyright Licensing Agency Rapid Clearance Service (CLARCS), 90 Tottenham Court Road, London W1P 0LP, UK; phone: (+44) 171 631 5555; fax: (+44) 171 631 5500. Other countries may have a local reprographic rights agency for payments.

Derivative Works
Tables of contents may be reproduced for internal circulation, but permission of Elsevier Science is required for external resale or distribution of such material.
Permission of the Publisher is required for all other derivative works, including compilations and translations.

Electronic Storage or Usage
Permission of the Publisher is required to store or use electronically any material contained in this work, including any chapter or part of a chapter.

Except as outlined above, no part of this work may be reproduced, stored in a retrieval system or transmitted in any form or by any means, electronic, mechanical, photocopying, recording or otherwise, without prior written permission of the Publisher.
Address permissions requests to: Elsevier Science Rights & Permissions Department, at the mail, fax and e-mail addresses noted above.

Notice
No responsibility is assumed by the Publisher for any injury and/or damage to persons or property as a matter of products liability, negligence or otherwise, or from any use or operation of any methods, products, instructions or ideas contained in the material herein. Because of rapid advances in the medical sciences, in particular, independent verification of diagnoses and drug dosages should be made.

First edition 1999

Library of Congress Cataloging-in-Publication Data

Reymont, R. A.
    Aspects of multivariate statistical analysis in geology / Richard A. Reyment and Enrico Savazzi.—1st ed.
        p. cm.
    Includes bibliographical references.
    ISBN 0-444-82568-1
        1. Geology–Statistical methods. 2. Multivariate analysis. I. Title. II. Savazzi, Enrico.
QE33.2.S82 R49 1999
550'.1'519535 21–dc21   99-044753

The paper used in this publication meets the requirements of ANSI/NISO Z39.48-1992 (Permanence of Paper).
Printed in The Netherlands.

Contents

Preface		vii
1	Introduction	1
2	*Graph Server* and *Graph Wizard*	39
3	Methods for analysing a sample drawn from a single population	103
4	Comparing samples from two populations: the discriminant function	155
5	Analysis of several groups: canonical variate analysis	173
6	Correlating between sets	211
7	Some problems in petrology and geochemistry	225
8	Miscellaneous examples	239
Glossary of computer program procedures		271
References		275
Index		281
Contents of accompanying Compact Disk		ix

Preface

Multivariate statistical methods have become commonplace in the Earth Sciences. What was once an exclusive area of activity is now within the reach of Everyman, owing to the ubiquitousness of mini-computers and the ready availability of software for doing the computing. In the days when one was required to do one's own programming, it was necessary to acquire considerable proficiency in linear algebra and one or more programming languages. Today, the vast majority of the people who use multivariate methods to analyse geological data have little or no idea of the matrix operations underlying a particular method, nor, for that matter, what the program is actually supposed to be doing. This situation can be both good and bad. It can do no harm if everything goes according to schedule, the program being used is competently constructed, which, alas, is far from being the general case, and there are no strong deviations from standard statistical theory in the data under examination. It is bad if the data do not fit the theoretical requirements of a particular method and even worse if the method of computation used is inappropriate. It is an inescapable and sad fact of life that much geological and biological material deviates in some manner or other from the theoretical requirements of a multivariate statistical procedure. The immediate relevance of this observation is that there are many sources of error in doing an analysis of geological data by means of standard statistical software.

The spread of multivariate statistics in the Natural Sciences has, therefore, taken place at a cost – the risk of doing something quite wrong and yet never knowing that a mistake can have been committed, or worse, that a blunder is even possible.

Books on multivariate statistics aimed at all levels of sophistication abound, from abstruse algebraically loaded treatises, through practically oriented texts, to volumes of computing recipes such as are profusely available for biologists. The special justification for this book is that it is concerned with the elementary consideration of the special types of multidimensional problems that occur in Geology and which are never, or are only summarily, considered in other places (e.g., textbooks dealing with multivariate statistics) and which cannot always be correctly analysed by commercially available software. There is an attached compact disk of compiled programs and trial data for doing the most commonly occurring computations, the files for running *Graph Server* and a file summarizing the steps involved in activating the various routines, but we lay no claims to perfection nor to elegance in the appearance of the computational output. There will be a www-site at the State

University of New York at Stony Brook for updates and corrections to the contents of the CD, thanks to the generosity of Professor F. James Rohlf. In two recent multivariate texts of RAR, the accompanying programs are written in copyrighted code, which means that the user is required to acquire the means of accessing this code. We have desisted from this practice, since it would defeat the main practical purpose of the book. We supply our own compiled FORTRAN and C programs with instructions for entering data; each method is illustrated by one or more sets of observations typical for the class of problem treated with an emphasis on the peculiarities of geological data. The programs have been constructed outgoing from our own research commitments. Note, however, it is not our intention to provide a self-contained, hierarchical system such as offered by F. James Rohlf's *NTSYSpc*. The compiled programs have mainly a didactic purpose – a simple means of illustrating the ideas expressed in the text. The important aspect of graphical presentation has been the province of Enrico Savazzi, whose new language *Graph Server (GS)* for displaying plots forms an integral part of the enterprise.

We have not provided more than a few introductory notes on the elements of matrix algebra, deeming that the basic manipulations required for being able to use the primer can be most effectively introduced at the appropriate points in the text. References to introductory manuals presenting the elements of linear algebra are provided wherever we have thought it necessary.

The idea for writing this introduction for geologists stems from more than 35 years of experience of RAR in teaching statistics to geologists and biologists in many parts of the world. This accumulated experience has convinced us that no matter how well people seem to be grasping a course in multivariate applications, and despite a maximum of teaching effort, the number of the participants who will really stay with the subject is small indeed. This is no reflection on the value of the discipline, but rather an indication of the difficulty experienced by the tyro in understanding what can be asked of statistical methods and how powerful multivariate methods are when applied in an appropriate manner. Frustrations occasioned by the incorrect use of techniques, including the "loyalty syndrome" with respect to a particular one (e.g. Correspondence Analysis among francophones) is a major source of disaffection. The main reason for this unfortunate situation is that when the average student has been cast out to swim on his own he will drown unless he has a lifebuoy with him. Only time can tell if this modest text is that lifesaver.

We wish to make it clear that the primer is not concerned with multivariate modelling of geological processes, regionalized variables, etc. Cluster-analysis as a specially delimited topic is likewise not taken up, notwithstanding that some of the techniques that have come to be associated with the concept of "clustering" are made use of – for example, the minimum spanning tree, similarity coefficients and Q-mode latent roots and vectors. (An easy introduction to clustering analysis is available in the book by Everitt (1974).) It is solely concerned with the simple application of standard methods of multivariate statistical analysis to geological data in the form of *arrays of measurements and compositional data* (e.g. chemical determinations). Consequently, we have not taken up advanced special methods of multivariate analysis such as M-estimators, robust methods, nor such interesting though relatively difficult procedures as the generalization of "biplots", notwithstanding that biplots are doubtless destined to play an increasingly important role in the future (cf. Gower and Hand, 1996). The full implementation of that subject is, moreover, still in the relative early phases of development. Where desirable, we provide references to more advanced statistical texts.

We are grateful to Professor John C. Davis, Geological Survey of Kansas and University of Kansas for a valuable and well reasoned criticism of the first version of the text from the geological standpoint. Professor John Aitchison, Department of Statistics, University of Glasgow, is thanked for precious advice on aspects of the analysis of compositions in connexion with a very early version of some of the chapters of the text as well as the current one. Dr. Allan Gordon, Department of Mathematics and Statistics, University of St. Andrews, kindly read the entire manuscript and furnished us with many thoughtful suggestions for improvements. Dr. Vera Pawlowsky-Glahn, Universidad Politécnica de Cataluña, Barcelona likewise read the entire text from the geomathematician's viewpoint and provided us with valuable suggestions and advice. For answers to various questions from Professor Leslie F. Marcus, Queen's College, New York and Professor F. James Rohlf, Department of Ecology and Evolution, State University of New York at Stony Brook, we are thankful. Professor Rohlf and SUNY are also thanked for making forthcoming updates available via the Internet. The updates will be made available by F(ile) T(ransfer) P(rotocol) at *life.bio.sunysb.edu/morphmet*.

We are well aware that some will react against the mode of presentation we adopt here, regarding it as "condescending", "non-academic" (because of the use of the first person rather than "the present authors" or the like, or the occasional use of phrases more appropriate to the spoken language than the written). This has been done with a definite purpose in mind. Multivariate geostatistics is not generally perceived as being an enthralling subject. A heavy "nuts-and-bolts" mode of presentation would have done little to help better matters.

The compact disk accompanying the primer contains the computer programs and teaching sets of data in two sub-directories, an HTML file explaining their use (being a summary of what is said in the main text), the files needed for using *Graph Server* and some files containing general information in a separate directory.

Finally, we wish to stress that this is a text for the IBM Personal Computer system (and clones). We have no plans for releasing a version for Macintosh machines. However, any Mac-adept with programming skills should be able to produce his own set of programs from the information provided in the text.

Uppsala and Stockholm, June 1999

Richard A. Reyment
Department of Palaeozoology
Swedish Museum of Natural History
Box 50007
10405 Stockholm
Sweden

Enrico Savazzi
Hagelgränd 8
75646 Uppsala
Sweden

Chapter 1

Introduction

AIMS AND SCOPE OF THE BOOK

In today's Geology, there is no shortage of quantitative information. Almost any kind of activity in the Earth Sciences generates data, often in quantities that surpass any possibilities of comprehension by inspection. A vast amount of the data accumulated is, however, never used for any meaningful purpose, being acquired automatically as the outcome of an administrative decision in the past, the reasons for which may have long lapsed into obscurity.

This text aims at introducing the geologist concerned with analysing multivariate observations to the appropriate methods of statistical analysis, at the introductory level. Our purpose is to demonstrate some of the more prominent multivariate statistical techniques for extracting the main features of relationships submerged in the data. We have not provided extensive background development of the methods, having, instead, made reference to standard statistical texts in which the algebraic aspects are given. In keeping with the spirit of the work, we have given questions dealing with deviations from the normal statistical situation particular prominence. Such data abound in the Geosciences, but are less common in biology and other areas of the natural sciences. This particular aspect of geological data is the most serious and common cause of flawed results in publications in the Earth Sciences.

A very large part of the multivariate data accumulated in geological work derives from analyses of rocks of various origins and complexity. These are usually chemical in nature, but sedimentology is also a frequent source of multidimensional observations. Let us consider some kinds of data encountered by geologists:

1 Chemical analyses of rocks.
2 Petrographical determinations of rock compositions.
3 Grain-size classes of sediments.
4 Frequencies of fossil species.
5 Frequencies of mineral species in a sediment.

6 Content of ore minerals in a mining sample.
7 Observations on physical properties of the crust.
8 Measurements made on fossils, pebbles, mineral grains.

Six of these categories have a property in common that is not shared by the seventh and eighth. All are multidimensional, but the row-entries for the first six sum to a constant. This is the principal feature of much multivariate data in Geology, namely, that the data-matrix is *constrained*, or closed, so as to have rows that sum to the same constant. It may not be immediately obvious to you what this ingredient of "closure" implies for statistical analysis. Suffice it here to say that constrained data must be treated in a manner that is different from what pertains for "usual" data. The sixth category is more elusive. The components of ore in a mining sample are not evidently constrained. However, the content of minerals is always expressed in relation to some measure of weight or volume, and it is this that imposes the constraint; the data are in terms of relative, not absolute, magnitudes and hence with ratios of components. The same situation arises in, for example, pollen analysis where counts on frequencies of species are made on samples of constant weight, constant volume, constant total of grains of all species counted, or samples that are reduced, by a simple division, to a constant standard of reference.

BASIC COMPUTING REQUIREMENTS

We have endeavoured to make our text "transparent", yet self-contained. There are certain technical requirements for using the text to your best advantage. Those that seem important to us are listed below. Well aware of the extremely rapid development in the personal computer these days, it is not possible for us to foresee what the capacity of machinery will be even in the next six months.

1 You will need an IBM compatible personal computer configured to operate under *Windows NT*, *Windows 95*, or *Windows 98* in order to use the compiled programs and *Graph Server*. The programs **will not run** in earlier versions of Windows. The programs are 32 bit compilations which means that they will not run unless you have installed the appropriate operational medium, i.e., *Windows NT* or *Windows 95, 98*. This might seem to be an unfriendly act and we thought long and hard about the implications for the user. An undeniable advantage offered by 32-bit technique is that it greatly speeds up computations. Moreover, innovations in programming languages, such as C++ and FORTRAN 90, are developed so as to run exclusively in a 32-bit environment.

2 The machine will need to be of the category 386, or higher in order to cope with the computations and the requirements of the Microsoft FORTRAN 90 and Microsoft C++ compilers.

3 You will need to have a CD-disk-reader attached to your machine.

4 Your PC will need to have a so-called mathematical coprocessor for the programs to work. A "math-chip" is standard equipment on 486 machines, Pentium-processor machines, and higher.

5 For larger applications, you can run into trouble if you lack extended memory. For most calculations, 8 Mb of RAM will prove sufficient, albeit slow. However, for the graphical work, *Graph Server* requires more random access memory. We recommend at least 32 Mb of RAM. (The computer programs were compiled and tested with 100 Mb of RAM.)

HIERARCHY OF METHODS

In the interests of simplicity, the methods are ordered according to the number of populations sampled in a particular connection and not according to theoretical statistical principles. The main structuring is as follows:

One Population:	Principal Component Analysis*
	Jack-knifed principal components
	Principal Coordinate Analysis*
	Q-R-mode combined analysis
	Multivariate Normality
Two Populations:	Linear Discriminant Function*
	Quadratic Discriminant Function*
	Generalized Statistical Distance*
	Hotelling's T^2*
Many Populations:	Canonical Variate Analysis*
	Multivariate analysis of variance*
	"Discriminant Coordinates"*
	Common Principal Component Analysis*
	Shrunken estimators for canonical vectors
Between-sets	Canonical Correlation Analysis*
Single Sample:	Multiple Regression

(Categories marked with an asterisk denote that the standard multivariate procedure is accompanied by a compositional counterpart.) Additionally, questions of accuracy, such as stability of estimations and robustness are reviewed where called for, which you will soon learn is quite often in much routine geological work. It should be mentioned that a grasp of the elements of *statistical inference* is required of the reader if the best use is to be made of this volume. We can recommend J. C. Davis (1988), which provides a coverage of most subjects of interest to geologists at a level easy of assimilation. A comprehensive reference of applied multivariate analysis is provided by the volume by Seber (1984). The text by Reyment and Jöreskog (1993) gives a detailed presentation of methods for analysing geological data drawn from a single statistical population. Of necessity, we have assumed the user to have a certain acquaintanceship with statistical principles. A handy reference for he who wishes to brush up on his knowledge is the book by Harris (1975). Although in some aspects a little outmoded, it contains much sensible information not always easily available in standard textbooks. The volume on multivariate trends edited by Krzanowski (1995) is another important reference for more advanced concepts. Mardia, Kent and Bibby (1979) contains many useful pieces of advice for the budding multivariate analyst. Lastly, an absolute necessity for the geochemist and the petrologist is the monograph by Aitchison (1986) on compositional data-analysis.

The last part of the text introduces a few "case-histories", that is, examples selected more or less randomly from the literature and which are here analysed, briefly, by appropriate methods in order to show the reader how a multivariate statistical analysis can be composed and to what extent the results yielded by the appropriate technique differ from or agree with those published in the original article, where applicable.

Anybody conversant with the theory and practice of multivariate statistics will quickly perceive that the subject matter of the text contains little that is new or controversial. All of the methods are well known and most of them have been used in geological research a great number of times. The only novel feature, perhaps, is that computing problems of particular relevance for geological situations occupy a prominent place. The appropriate methods for doing such analyses are generally not available elsewhere in standard "packages", many of which have been produced for applications in the Social Sciences and which, understandably, are not of unrestricted value in the Earth Sciences.

A knowledge of linear algebra will prove useful, if not enlightening for many applications. There are several books we can recommend. A very readable introduction is that of P. J. Davis (1965). Searle (1966) gives a well-illustrated account of biostatistically oriented matrix algebra. Reyment and Jöreskog (1993) contains a chapter on linear algebra (Chapter 2) illustrated in geological terms. For those with more advanced tastes, the two volumes by Gantmacher (1965) can be recommended. The Appendix in Anderson (1984) is another useful source of information. If you should want to go ahead and try your hand at programming matrix manipulations, we can personally strongly recommend Zurmühl (1964).

In practice, multivariate data in the Earth Sciences seldom conform with neat theoretical requirements. Whereas it may not be possible to provide tailored solutions to every kind of problem that can occur, much can be done to improve the scientific quality of an analysis by assessing and 'ferreting out' divergent specimens, such as are liable to distort the statistical interpretation of results, and to recognize the most efficient means of dealing with a particular problem. Divergent specimens are not necessarily wrong. There is a dichotomy of reasoning involved here in that an atypical observation may deviate in the statistical sense by differing markedly from its fellows, but it may be perfectly good in the geological or biological sense. For example, measurements on good specimens of fossil (and living) organisms may need special treatment owing to the occurrence of shape-polymorphism.

For practical purposes, statistical methods may be considered to fall into one of the two categories, *descriptive* and *inferential*. Descriptive techniques are those that make statements about a given set of data, whether it be a single sample or a complete population, with the end in view of providing data analyses or graphical representations of the data. The inferential category encompasses procedures that take the analysis a stage further by making statements about the population from which the sample was drawn and in connection with which, probability plays a prominent role. For the most part, we shall here be concerned with descriptive aspects of statistical analysis, such as occur in multivariate exploratory data analysis. It is, nonetheless, important to keep the above distinctions in mind. Many examples of geological data analysis cannot be extrapolated to the general case in the manner in which the analyses were done. Instances of this "fixed-mode category" will be encountered further on (cf. Reyment and Jöreskog, 1993).

Some examples of problems in statistical geology

1 A particular kind of statistical constraint is provided by orientational data: dips, strikes, palaeocurrents, etc. The most common way of expressing such observations is by means of a so-called Rose Diagram, a sort of circular histogram (Hoorn, 1994, p. 10). Much more information can be extracted if such material be subjected to appropriate statistical analysis (cf. Reyment, 1971). Hoorn's study is for fluviatile environments in the Amazonas Basin. One of the diagrams shows pronounced bimodality and the data would have profited from numerical analysis.

The same paper brings to light one of the more common fallacies occurring in geological publications. A strong product-moment correlation is reported for the number of taxa in relation to the sum of recovered pollen grains. Inspection of the accompanying figure (Hoorn, 1994, Fig. 15) indicates that this high correlation is largely due to the fact that the observations are heterogeneously distributed over two well-separated and disjunct fields. Such a circumstance will unfailingly give rise to spuriously high correlation coefficients.

2 Giresse et al. (1994) take up the interpretation of late Quaternary palaeo-environments in Lake Barombi Mbo (Cameroun). Correlations are computed for percentages, without regard for the constraint and ratios are plotted against a component of that ratio. This is common procedure in petrology but one that must be decried, despite its hoary status, not least if correlations between such composites are computed (a part is being correlated with itself). Pearson (1897) in his classical paper on spurious correlations warned against this type of procedure.

Rao and Jayawardane (1994) analysed major minerals with respect to elemental and isotopic compositions in modern temperate shelf carbonates in Eastern Tasmania, Australia. Their data consist of proportions for which correlations were calculated and unspecified "factor analysis" was carried out without taking the constraint into consideration. Some of the results thus invite suspicion, for example, those reported in their Figs. 4, 5, 6 and 7.

3   Tribollivard et al. (1994) presented results obtained for the study of organically rich cycles in the Kimmeridge Clay of Yorkshire in relation to conditions of oxidation and reduction and determined for an Upper Cycle and a Lower Cycle. Again, we meet compositional data (the determinations are in weight percent) which were analysed without recourse to the appropriate statistical model (cf. Aitchison, 1986). This is a didactically instructive case and it will be reviewed in some detail. The **parts** considered in the present connection are SiO_2 (1), Al_2O_3 (2), Fe (3), S (4), Total Organic Content (5) and $CaCO_3$ (6). The observations made on these oxides are termed the **components**, that is, the numerical proportions in which individual parts occur. The correlation between SiO_2 and Al_2O_3, cited in the text, is roughly the same for both the constrained version and the "open" version and it might be argued that there is little to be gained by being correct. This is, however, fortuitous. Table 1 contains latent vectors for the Lower Cycle obtained by the appropriate model (using the centred log-ratio covariance matrix) and contrasted with the inappropriate one obtained from the raw data via the product-moment correlation coefficient. The strong differences between the two categories are in part due to heterogeneity in the data-set for the Lower Cycle, which was one of the points arrived at by a graphical appraisal by Tribollivard et al. (1994). The more homogeneous Upper Cycle displays, however, marked divergences as well between the results obtained by the inappropriate and appropriate procedures.

4   Series in time occur frequently in geological work, particularly those that deal with stratigraphical material. It is commonplace to find comparisons of curves used to illustrate regional correlations, for example, comparisons of pollen sequences, borehole logs, and like observations. "Eye-balling" such graphs is not a trustworthy

TABLE 1

Latent vectors for the Lower Cycle correlations (Tribollivard et al., 1994)

Parts	First latent vector		Second latent vector	
	Constrained version	Constraint ignored version	Constrained version	Constraint ignored version
SiO_2	0.42	−0.50	−0.45	0.14
Al_2O_3	0.40	−0.47	−0.51	0.33
Fe	0.40	0.41	0.25	0.11
S	−0.45	0.38	0.03	0.50
Total organic	−0.44	0.31	−0.20	0.56
$CaCO_3$	0.33	0.35	0.65	−0.54

approach, and the often-invoked method of cross-correlation is not a sound procedure in geological work (the technique comes from econometrics and requires exactly specified time-intervals). Reyment (1991) promoted the method of "slotting" which provides a reliable basis for correlating series in time. Moreover, Bookstein (1987) pointed out the important fact that a single "evolutionary sequence" cannot be claimed to represent a genuine trend unless certain specific requirements are met (Feller's coin-tossing paradox) because it could well have arisen as a simple random walk in time. Bookstein and Reyment (1992) give a further example of trend versus random walk in the analysis of geological data.

Remarks: These few, brief reflections serve to show that the correct use of statistics, particularly multivariate statistics, in the Earth Sciences requires a level of proficiency and technical understanding that is seldom to be found among geoscientists. A major contributing cause to this lies with moribund university curricula and the internationally prevailing dearth of competent instruction in statistical methods in the Earth Sciences. It is therefore gladdening to observe that the *International Association for Mathematical Geology* has recently (1997), albeit belatedly, realized that something must be done and has initiated a series of excellent international symposia on the subject of appropriate methods of analysing compositional data. It has taken a very long time to get the geological community to begin to accept that vast sums of money are being expended on research projects, the statistical treatment of which is gravely flawed – it is, after all, almost 20 years now since Professor John Aitchison took up the subject in detail!

NOTES ON THE RELATIONSHIPS BETWEEN MULTIVARIATE METHODS

Multivariate methods should not be regarded as a heterogeneous collection of isolated procedures. One may gain that impression, however, owing to the fact that they have been developed as specialized answers to specific practical problems (Reyment, 1996). It is only fairly recent practice to present the multivariate statistical corpus in a wider context.

Many of the most frequently used methods were developed in the 1930s. There were many controversies at the time in connection with some of them, the strangest of which was that concerning the Pearsonian "Coefficient of Racial Likeness", the Mahalanobis generalized statistical distance, D^2, and Hotelling's T^2. It was finally recognized that Pearson's coefficient is just Mahalanobis' distance for zero correlations and that Hotelling's measure is the conversion that expresses the statistical significance of a generalized distance, that is, can be used to test hypotheses about a generalized distance.

Very many practical problems are concerned with analysing a single sample drawn from a *population* (which is defined as the total number of specimens in existence of the category). The early statisticians had to rely on very large samples for their work in order to approximate the properties of the theoretical population. An important development came in 1908 when W. S. Gosset, a statistician with Guinness Breweries, Dublin, put forward his celebrated "Student's" t-test, for comparing two univariate means on small samples.

The generalization of the algebra appropriate for univariate samples is achieved via *linear algebra*, the algebra of two or more variables considered simultaneously.

Much routine multivariate statistical work consists of some form *of principal component analysis*. The original idea for this came from Karl Pearson who, at the turn of the century, was concerned with lines and planes of closest fit to points in space. He was led to the problem though dissatisfaction with the concept of dependent and independent variables in regression analysis. Pearson's argument was developed in geometric terms with reference to the principal axes of an ellipsoid (which represent the principal components) and the regression lines fitted to the points encompassed by the hull of the ellipsoid.

Principal component analysis is concerned with the space spanned by the p variables; it is referred to as an R-mode method (from the use in psychometry of the correlation matrix as the starting point for factor analysis – the common symbol for the sample correlation matrix is **R**). The need to examine data from another point of entry was felt at a fairly early stage in psychometric work. The way in which this was solved at first was by computing the principal components for the space spanned by the N specimens of the sample (which thus took over the role of the variables of the usual mode of analysis). This procedure became known as *Q-mode factor analysis* and, indeed, many special techniques were borrowed from psychometric factor analysis in the R-mode, albeit clandestinely. The methodology was placed on a sound footing by J. C. Gower (1966) under the name of *Principal Coordinate Analysis* (also termed principal coordinates analysis – we prefer the term without an 's' for grammatical reasons). He also put forward a universal *similarity coefficient* (Gower, 1971) which encompasses quantitative, qualitative and dichotomous variables.

Many applications in Geology and Biology perpetuate a misconception when they speak of *Factor Analysis*. The mathematical model appropriate to true factor analysis is almost exclusively the domain of psychometrics. That which is called Factor Analysis in the natural sciences is in reality an extension of principal components upon which some of the techniques of the psychometricians have been superimposed.

In summary, it may be said that in terms of the *Data Matrix*, principal components treat the columns (the variables) and principal coordinates the rows (the number of specimens on which the variables have been measured). There is an algebraic relationship between R-mode and Q-mode, based on covariances, which is expressed by the *singular value decomposition of a matrix*. This is known as determining the "basic structure of a matrix" in psychometrics, where the technique first achieved prominence in statistical work (although it has long been known to mathematical physicists). By means of a scaling procedure, R- and Q-mode representations of a data matrix of frequencies can be depicted on the same figure.

Such figures are referred to as *biplots*, the best known of which is referred to as Correspondence Analysis (Benzécri, 1973), which is a special case of the Gabriel biplot.

What can be done if each set of observations consists of, say, measurements on one kind of property and a second set on another kind? Such data arise in environmetrics where one set may consist of measurements on the morphological variation of a species and the second set might be a set of chemical determinations. The appropriate technique for such data is known as *Canonical Correlation Analysis*, whereby the one set is correlated with the second one. Canonical correlation can be generalized to more than two sets. When this is done, and the scaling procedures of correspondence analysis are incorporated, the ecologically useful method of *Canonical Community Analysis* results (ter Braak, 1987). We introduce canonical correlation in the following, but not canonical community analysis, largely because of its predominantly environmetric flavour. There is a well constructed computer program *CANOCO* available from Dr. C. J. ter Braak (Wageningen) that does the computations to which we refer anybody who becomes involved in advanced ecological research.

If the specimens have been sampled from more than one population, several avenues of interest lie open. In the case of **Two Populations**, the *Linear Discriminant Function* of R. A. Fisher is a fitting procedure. In its original form, the discriminant function was designed for sorting objects, with a minimum of error, according to the original populations from which they came, under the necessary assumption that they really derived from one of two populations. The method was first applied to anthropometric data and then to a problem in plant taxonomy. Just prior to Fisher's discovery, P. C. Mahalanobis invented a measure of multivariate statistical distance for anthropometric work in India, the *Generalized Statistical Distance*, known widely as the D^2. The significance of the generalized statistical distance is assessed by a variance ratio, known as Hotelling's T^2, which was introduced as a type of multivariate generalization of Student's t.

When there are more than two populations, the generalization takes several forms, often united under the name *Canonical Variate Analysis*. A complete such analysis incorporates:

1   A comparison of the multivariate means by a generalized one-way analysis of variance – *MANOVA*, the acronym for multivariate analysis of variance. This test is to ascertain whether the k samples could have come from one and the same population.
2   A set of discriminant functions for allocating specimens, with a minimum of risk, to the appropriate populations. The computation of this function proceeds by extracting latent roots and vectors and is therefore superficially similar to principal component analysis. For just two populations, the same discriminant function as would have been yielded by the two-sample computations (which do not resort to latent roots and vectors) is yielded.

3 There is usually a graphical display of discriminant function scores. One says that the scores are *ordinated*. This is often considered to be the main reason for wanting to do the analysis, particularly at the exploratory level.

4 A complete canonical variate study should also include the generalized distances between samples, the associated values of the Hotelling T^2, and the corresponding variance ratios.

The correct use of multivariate statistical procedures should have as its guiding light the selection of the appropriate technique or techniques for solving a particular problem. All too often, it is the ready availability of some method or other that generates a search for research materials that are considered to fit the assumptions of that method.

THE MULTIVARIATE SAMPLE

Some basic concepts and definitions

An array of data consisting of N specimens on which p characters have been measured is called a *data-matrix*, denoted here as **X**. Thus, each row of **X** is constituted by a vector of observations containing p components. You will seldom find this convention in strictly statistical texts where the discussion is couched in terms of a p-dimensional vector of random observations. The data-matrix is, however, a very convenient concept in applied multivariate analysis, including geostatistical practice.

When the information contained in the matrix is analysed so that the p variables are compared to each other, which is the most common approach, the analysis is said to be in the R-mode. If the alternative way is chosen, namely, to analyse the data-matrix so that one specimen is compared to another, the analysis is said to be in the Q-mode. In other words, R-mode applies to the treatment of the p columns of the data-matrix and Q-mode is specific to its N rows. As already mentioned, the letter R comes from the standard representation of the sample correlation matrix, which forms the foundation of many analyses in psychometrics. The letter Q has no other significance than that it precedes R in the alphabet. This usage also comes from the field of Psychometrics. It is possible to unite both modes into a single graphical representation, referred to as a Q-R-mode figure, although not without certain reservations, according to Gabriel (1995a, 1995b), the inventor of the Biplot. There seems to be some confusion about how the R-Q-mode design is to be employed. For example, Laenen et al. (1997) used what they referred to as a Q-R-mode factor analysis, without supplying adequate information concerning which Q-R-technique was used. It is clear that factor analysis was not involved, though, possibly, correspondence analysis was meant. There is an additional issue arising from the analysis of a table of frequencies, namely, that ratios of parts were used in place of the parts themselves; this is a practice that is proscribed by professional statisticians, dating from the early works of Karl Pearson and Francis Galton a hundred years ago. The reason is that false correlations easily arise. Karl Pearson (1897) was most emphatic about the risk of computing spurious correlations and, in particular, the need to avoid attempts at interpreting correlations *between ratios the numerators and denominators of which contain common parts*.

The **mean vector** is the vector composed of the means of each of the p variables. It is then the sum of all the rows of that data-matrix, **X**, divided by N. It forms the centroid or barycentre of an empirical distribution.

The **covariance matrix** is a square $p \times p$ array formed from the variances and covariances of the p variables. The sample covariance matrix is usually written as **S**. It has the p variances ranged along its diagonal, the s_{ii} ($i = 1, \ldots, p$) and the covariances, s_{ij} in the off-diagonal positions. That is,

$$s_{ij} = s_{ji} \, (i = 1, \ldots, p; \, j = 1, \ldots, p)$$

which means that the matrix is a square symmetric matrix. The covariance matrix can be thought of as being a generalization of the variance of univariate statistics. By way of comparison, the data-matrix is neither symmetric nor square, other than by pure chance.

The *matrix of correlations* corresponding to **S** is written **R**. It has ones down the diagonal and correlations r_{ij} ($i \neq j$) in the off-diagonal positions. It is, then, also a square symmetric matrix.

These quantities enter into almost all of the methods considered in the following pages. Before going any further, we should mention two conventions. The first concerns Greek letters for parameters. This was introduced by R. A. Fisher in order to make it easy to distinguish between theoretical, population quantities (Greek letters) and their sample estimates (Roman letters). Most statisticians tend to follow this but not even Fisher himself could be relied on to do so always. The second convention concerns the use of bold lettering. It has become more and more widespread, although by no means universal, to use bold type to denote vectors (lower case) and matrices (upper case) and to reserve normal italicized type for scalars.

In addition to the covariance and correlation matrices, another type of representation of covariation appears in some connections, namely, various kinds of *Association Matrices*. These are square symmetric matrices, the elements of which express degrees of "association" or concordance in pairs of categories. In the present text, such matrices are employed in the Q-mode method of principal coordinates (Gower, 1966). It is quite possible to regard the correlation matrix as a particular variety of an array of associations. In some fields of quantitative work in biology and geology, particularly those that are concerned with what is known as "Classification", the study of association matrices is given much prominence. A handy reference is the book by Gordon (1981).

We shall start by introducing a simple program for computing a covariance matrix, its correlation matrix, and a mean vector. It is called *covmat*. Type

covmat

at the *DOS*-prompt. The program will ask for an input file. This is the data-matrix, specified as follows, and provided on the disk under the name of *haiticvt.dat*, a set of chemical analyses for the celebrated Cretaceous/Tertiary bolide glass impact near Haiti, given great prominence in the ongoing speculation about the demise of the dinosaurs, ammonites, belemnites, etc. (Sigurdsson et al., 1991; Reyment and Jöreskog, 1993, p. 281). The chemical parts are the oxides SiO_2, Al_2O_3, MgO, CaO, Na_2O, K_2O, and SO_3. Note, that we say "parts", not "variables". The reason for making this distinction is taken up later on. Suffice it here to say that compositions are not variables in the statistical sense of the term. Moreover, compositions are concerned with relative, not absolute magnitudes.

Line 1: N, p

N is the number of columns (i.e., the number of objects); the comma here is important – it is a requirement of the *C*-language (there is no corresponding punctuation mark in FORTRAN)
p is the number of variables

lines 2 and following, the data-matrix.

Type in *haiticvt.dat*

then press "Enter".

The program provides firstly the covariance matrix, then the mean vector, with each element numbered, and finally the correlation matrix of the data-matrix stored in the file *haiticvt.dat*. You might find this little routine useful for doing simple preparatory work.

Simple arithmetic operations with matrices

Addition and subtraction are carried out in the same manner as in usual or scalar arithmetic. The only difference is that there are more quantities, $p \times p$, to be precise, where p denotes the dimensionality of the matrix.

Multiplication and inversion are, however, not direct equivalents of their scalar counterparts.

The formula for multiplying matrices $\mathbf{A}_{ik}$ (*i* rows and *k* columns) and $\mathbf{B}_{kj}$ (*k* rows and *j* columns) is

$$\mathbf{C}_{ij} = \mathbf{A}_{ik}\mathbf{B}_{kj}.$$

This is a relatively simple calculation. Matrix inversion is a more complicated procedure.

You can gain an idea of these basic matrix operations by means of the two small programs provided. The first is called *matops*. Type *matops* at the *DOS*-prompt. You will be asked to provide a matrix for analysis. Start with the file called *matops.dat*. There are further examples in data-files *matops2.dat*, *matops3.dat* and *matops4.dat*. These are matrices that are added, subtracted and multiplied and the results listed. Some of the matrices are square and can be added, subtracted and multiplied. Those pairs that are not square and have different numbers of rows can be multiplied in the examples provided.

For matrix inversion, use the program *matinv*. Run the file *matinv1.dat*, a correlation matrix, then the file *matinv2.dat*. The second data-set is the inverse correlation matrix you have just used. Its inverse should (approximately) restore the original matrix. When you inspect the outputs, it will become obvious that inverting a matrix is not just a matter of writing down the reciprocal of each element for matrix inversion involves a recognition of interrelationships between elements.

This is all on the topic of elementary matrix manipulations for the moment. The subject is broached again in the introduction to the section on principal component analysis in Chapter 3.

Reification

One of the main interests in doing a multivariate analysis is to interpret the results in specific terms. This method of obtaining insight is called *reification* (which in the present connexion just means explaining and comes from the Latin root "res" = a thing) and it is more the domain of the scientist than the statistician, who cannot be expected to have an a priori notion of what is reasonable for a particular scientific problem. Reification is more an art than a skill and it usually requires considerable scientific talent to be able to unravel what the calculations have produced. It is also a source of dangerous conclusions and it is necessary to be aware of this and not try to squeeze out more "information" than can be reasonably yielded by a set of observations.

Kinds of variables

A few words need to be expended on the way in which multivariate data occur. The most common class of variables is the *continuous variable*, that is, one that can take all values over a certain range. Examples of continuous variables are lengths, distances, weights, temperatures, pressures, etc. A *discrete variable* is one that takes integer values such as, 1, 2, 3, 4 etc. (i.e. whole numbers). As an example of one type of this category, we can cite the number of spines on a fossil shell, the number of ribs on a bivalve, and the number of sedimentary bands in a formation. The *nominal scale* of measurements is one in which there are several mutually exclusive categories of equal rank. As an example, we can take 'type of rock', with mutually exclusive categories limestone, granite, shale, etc. Another variety of discrete observations is rank, or *ordinal*, data which arise when there is a hierarchy of states, such as in the grade of metamorphism, or Mohr's scale of hardness for minerals. Typical of such data is that the steps between successive states are not equal.

Binary variables are a special case in which the observations can take only one of two values 0 or 1, plus or minus, present or absent. This kind includes dichotomies such as males and females, presence or absence of a feature, such as the presence or absence of mineral species in a petrological sample, or of fossil species in a sediment, the presence or absence of an ornamental feature on a fossil.

Qualitative variables are sometimes of interest. These have no natural number associated with them, but can be arbitrarily coded in a manner amenable to data-analysis. Hues of soils and rock-types can be treated in this way, subjective estimates of degree of metamorphism, and so on.

Gordon (1981) has provided a concise and well presented review of the classes of variables in statistics to which we refer you for further remarks and references.

COMPOSITIONAL DATA

The most common type of observations occurring in the geosciences are compositions of various kinds for which reason we present a relatively detailed account of their properties. The significant aspect of compositions is that the data are in the form of frequencies, proportions or percentages, all of which have the common property that the rows of the data-matrix sum to the same constant. This may not strike you as being much of an obstacle, but rest assured, there is no other area of data-analysis in which more incorrect applications of statistical methods have been perpetrated, and not only by non-specialists. Until Aitchison (1986) monographed the algebra of the unit-sum constraint, the only avenues open to anyone wanting to try to get around the difficulty were to ignore it entirely, to wish it would go away, or to devise some totally inappropriate statistical analysis. In many cases, disastrous consequences have resulted. Surprisingly enough, even highly competent professional statisticians have done all the above listed wrong things, not just natural scientists. The otherwise so comprehensive volume on multivariate analysis by Jackson (1991) glosses over the entire problem posed by compositional data in just two pages!

"Closed data" require special methods for their correct analysis. Notwithstanding this sweeping declaration, you should be aware that the last word concerning the analysis of compositional data may not yet have been spoken. We shall try to demonstrate the motivation for saying this to you in a later section. A recent article by Barceló et al. (1996) casts light on some of the practical difficulties still remaining in the analysis of constrained observational vectors.

An awareness of the need for the appropriate way of dealing with compositional data is beginning to spread throughout the geological community, albeit at a gastropodous pace. There is widespread bewilderment evident in petrological publications at what the essential characteristics of compositional data-analysis are and how various challenges are to be met. Fortunately, in an increasing number of publications, the need for the log-ratio covariance matrix in the multivariate analyses of rock analyses is coming to the fore.

Definition of compositional data

In order to clarify matters, we review below the nature of compositional data. Geology does not have sole right to the problem of "closure" – examples abound in medical statistics, serology, psychometrics, ecology, zoology, sociometrics, economics, etc. We, as geologists, can, however, preen ourselves in the knowledge that we at least know about the problems involved and are starting to do something about them – other disciplines are still in a state of blissful ignorance. The study of compositions is essentially concerned with the relative magnitudes of *ingredients* rather than their absolute values, such as is the case for, say, measurements on some object such as a fossil specimen. These ingredients are **parts** and not variables in the real sense of that term in statistics. That is why chemical data were not cited above as being an example of "variables". Making this distinction may seem to be pure "hair-splitting", but a moment's reflection should convince you that there is a very real, and important, difference between variables and parts.

Any vector **x** with non-negative elements

$$x_1 + \ldots + x_D = 1 \tag{1:1}$$

is subject to the unit-sum constraint. This condition is referred to as being a composition **x** composed of D parts summing to 1. Obviously, the components of eqn. (1:1) cannot be independent since they sum to one.

As geologists, we meet such data in geochemistry, petroleum chemistry, sedimentology, rock-analysis, palynology, palaeoecology, oceanography, environmetrics, etc. In fact, it could almost be claimed that "closed data" are the most commonly occurring forms of measurements in general geology. Alluding to biology, we note that all serological data are of this kind, e.g. blood-group frequencies, observations on species occurrences, quantitative ecological observations, and many more. One hundred years ago, Karl Pearson and Francis Galton warned against attempting to interpret correlations between ratios, the numerators and denominators of which contain common parts. It should not be necessary to alert geologists to this danger today, but this is unfortunately not true and some petrologists still base much of their work on such fallaciously compounded data. For example, Noll et al. (1996) based an intricate succession of arguments concerning the role of hydrothermal fluids in the production of subduction zone magmas on bivariate plots involving ratios. Some of these plots feature values in which one axis consists of a second chemical element divided by the element forming the other axis. Other plots (Noll et al. 1996, p. 596) express bivariate regressions, and correlations, for ratios in which the same element enters into the "dependent" variable and the "independent" variable. It is difficult, yea perilous, to relate such analyses to statistically valid concepts.

The characteristic features of a compositional data-set are:

(a) each row of the data-matrix corresponds to a single specimen (i.e. rock sample); this is known as a *replicate* (= a single experimental or observational unit).

(b) each column of the data-matrix represents a single chemical element, a mineral species, in short, a **part**;

(c) each entry in the data-matrix is non-negative;

(d) each row of the data-matrix sums to 1 (proportions), respectively, 100 (percentages). (N.B. you will sometimes find another row-constant, owing to some manipulation or other on the part of the analyst);

(e) correlation coefficients change if one of the variables is removed from the data-matrix and the rows made to sum to 100 again. This is the property of variable-dependent correlation. The same effect is also produced if a new component is added to the study.

Property (e) provides part of the key to the predicament attendant on compositional data. By way of comparison, correlations computed for non-compositional data-matrices are invariant to the number of variables included in the matrix. That is, if you delete one or more variables from the data-set, this action has no effect on the correlations between the remaining variables. Concerned petrologists have worried over ways of attacking the constant-sum problems and many suggestions, all fallacious, abound in the literature. The most recent try at finding a simple solution is the constant-weight artifice.

The constant weight stratagem

Whitten (1995) advocates the use of constant-weight samples in order to "break out of the constraint". By realizing this intention, the rows of a data-matrix of chemical parts will not necessarily sum to a constant. There is, however, a constraint resident in the data in that the parts are interrelated in the same manner as if the observations were expressed as percentages of a whole. This will only be of immediate consequence for a particular rock-analysis if a subcomposition is selected subsequent to the construction of the data-set. Removal of one or more parts alters the weight of the sample (row) in an unforeseeable way and hence the relationships between parts. The constant-weight stratagem can be considered as a **fixed-mode** procedure (Reyment and Jöreskog, 1993) insofar as the results obtained for a certain data-matrix apply for that set of observations alone, although, without 'rending asunder' the constraining relationship in any wise. Aitchison (1997) has pointed out several fallacies in the constant-weight stratagem. Statistical extrapolations are normally not permissible in the fixed mode other than on an ad hoc basis. Whitten (1993) advocated g/100cc of rock rather than weight-percent for studying spatial chemical variability. The volumes were obtained by the familiar $V = M/d$ relationship where M is the weight percentage and d denotes the whole-rock specific gravity; it was, however, pointed out that whole-rock specific gravities are notoriously fickle.

A point that is overlooked consistently by petrologists is that their data were not singled out to be pilloried exclusively by Aitchison (1986). Such observations belong to a much wider world of compositional data which encompasses not only rock analyses, but such diverse material as analytical chemistry, serological determinations, time and motion study analyses, analytical petrochemistry, econometrics, etc. The vernacularized and, frequently, emotionally charged terminology used in some geological literature is indeed unfortunate and a source of more than a mite of misunderstanding.

The simplex

A restricted part of real space, the *simplex*, constitutes the basic concept for the treatment of compositional data. The essential point you need to grasp at this stage is that although the vector **x** in (1:1) consists of D parts, the composition it represents is **completely specified** by the d components of a d-part subvector, defined as $d = D - 1$. Hence,

$$x_D = 1 - x_1 - \ldots - x_d. \tag{1:2}$$

A D-part composition is therefore, to all intents and purposes, a d-dimensional vector. If you know the sizes of these d parts, x_D can be found by simple subtraction from the row-constant. The concept of the space of compositional data can then be simply defined as the d-dimensional simplex embedded in D-dimensional space.

A convenient working definition of the simplex can be made in the following words: The d-dimensional *simplex* embedded in D-dimensional real space is the set defined by

$$\mathbf{S}^d = \{(x_1, \ldots, x_D): x_1 >, \ldots, x_D > 0; x_1 + \ldots + x_D = 1\}.$$

In this manner, $\mathbf{S}^d$ is defined as being embedded in a real space of higher dimension, $\mathbf{R}^D$.

This condition can be exemplified by a simple geologically relevant illustration, the compositional triangle. $\mathbf{S}^2$ is an equilateral triangle, a two-dimensional subset, within $\mathbf{R}^3$. It is the triangle associated with the familiar ternary diagram of petrology, sedimentology and physical chemistry.

There is, additionally, the problem of the **graphical display** of constrained data. Aitchison (1986. p. 51) summarized some of these and noted that the Harker diagram of petrology, in which SiO_2 is chosen to be the component against which all other major oxides are compared, is a case of this, particularly if interpretations in the usual framework of correlation and regression are attempted.

Bases and compositions

A *basis* $\mathbf{w}$ of D parts is a $D \times 1$ vector of positive components $(w_1, \ldots, w_D)$, all on the same scale. The **constraining operator** $\mathbb{C}$ of Aitchison (1986) offers a convenient means of representing the transformation between bases and compositions. It transforms each vector $\mathbf{w}$ of D positive components into a unit sum vector $\mathbf{w}/\mathbf{j}^T\mathbf{w}$, where $\mathbf{j}$ denotes the **unit vector** (i.e. a vector the components of which are all ones).

Every basis $\mathbf{w}$ has a unique size, namely, the sum of its components:

$$t = (w_1 + \ldots + w_D) = \mathbf{j}^T\mathbf{w} \tag{1:3}$$

and

composition $\mathbf{x} = \mathbb{C}(\mathbf{w}) = \mathbf{w}/t.$

This is a very neat way of expressing the relationship between base and composition.

Subcompositions and amalgamation

1 Subcompositions

The formation of a subcomposition is not merely a matter of deleting a part from each composition in a manner analogous to what is appropriate for usual variables (e.g. deleting one or more distances from a set of craniometric traits). The process of selection of a particular subcomposition may be conveniently described in the following terms. If S is any subset of the parts $1, \ldots, D$ of a D-part composition $\mathbf{x}$, and $\mathbf{x}_S$ is the subvector formed from the corresponding components of $\mathbf{x}$, then $C(\mathbf{x}_S)$ is called the subcomposition of the parts S. We can illustrate the pithiness of this terse formulation by considering a simple example for 5 parts, $\mathbf{x} = (x_1, \ldots, x_5)$ from which parts x_1, x_4, x_5 are selected to form a subcomposition.

$$(s_1, s_2, s_3) = C(x_1, x_4, x_5).$$

Geometrically, this is a transformation from the original sample space $\mathbf{S}^4$ to a new simplex $\mathbf{S}^2$. The pertinent matrix multiplication is (note the rôle of the ones in the $C \times D$ premultiplying matrix):

$$\begin{pmatrix} 1 & 0 & 0 & 0 & 0 \\ 0 & 0 & 0 & 1 & 0 \\ 0 & 0 & 0 & 0 & 1 \end{pmatrix} \begin{pmatrix} x_1 \\ x_2 \\ x_3 \\ x_4 \\ x_5 \end{pmatrix}$$

An important property of compositional data is that the *ratio* of any two components of a subcomposition is the same as the ratio of the corresponding two components in the full, original composition. Hence,

$$s_i/s_j = x_i/x_j,$$

which is the property of "preserved ratio relationships". We stress, again, a matter of computational logic here, namely, that if it is desired to reduce the dimensionality of an analysis, the reduction must be carried out by the appropriate procedure for subcompositions and not by the simple expedient of lopping off part of the data. Aitchison's (1997) enlightening discussion of this subject should make the rationale quite clear.

2 Amalgamation

Amalgamated data-sets are often of interest in geological work. The original data may, for example, be in the form of oxides of Na_2O, K_2O, Fe_2O_3, FeO, but subsequent interest is concerned with total alkalis and total iron-content, thus leading to an amalgamation of parts. *Amalgamation* is defined as follows. If the components of a D-part composition are separated into C ($<D$) mutually exclusive subsets and the parts constituting each subset are added together, the resulting C-part composition is termed an *amalgamation*. The addition of components is expressed by the matrix operation:

$$\mathbf{t}^{(C)} = \mathbf{A}\mathbf{x}^{(D)}.$$

An amalgamating matrix $\mathbf{A}$ is one of order $C \times D$ with D elements equal to 1 (one in each column) and at least one in each row; the remaining $(C-1)D$ elements are equal to zero.

Covariances and correlations in simplex space

The usual covariance matrix, when computed for a D-part composition, runs into interpretational difficulties. Some of these are:

1 Negative bias. The correlations are not free to range over the interval $(-1, +1)$. This becomes obvious in the case of two variables for which

$$\text{corr}(x_1, x_2) = -1.$$

Hence, the product-moment correlation coefficient is constrained to taking a specified value, which is, of course, quite unacceptable.

2 There is no relationship between the product-moment correlations of a subcomposition and those of the full composition. As the dimensionality of a subcomposition is decreased, so do the crude covariances (correlations) between two specific parts fluctuate in sign. This is hardly a useful property. The inadvertency noted here is in part due to the incoherency of the correlation coefficient in simplex space.

3 The way in which null-correlation is manifested is a further bugbear. A value of zero computed for the raw correlation coefficient of two parts of a composition is almost always an untrue representation of the real situation expressed by the appropriate correlation in simplex space. Aitchison (1986, p. 56) reviewed the various attempts that have been made to come to grips with the difficulty of null correlation, including the construction of "imaginary bases" (sometimes called the "open set"). This approach seems to have arisen by false analogy with true factor analysis. The "open set" is interpreted as being the "hidden" relationship between parts from which the observed compositions, the "closed set", could have arisen. This manipulation introduces a construction, the imaginary basis, which for truly compositional data does not exist. The whole idea of "opening up" a compositional relationship is based on a logical *non sequitur*, as shown by Aitchison (1986, p. 58). The concept of the zero correlation does not have the same meaning with respect to independence as is the case for "usual data".

Attempts have been made in the past, such as the Chayes–Kruskal method, to produce a null correlation by some kind of manipulative "opening procedure". However, the whole idea of null correlation for compositional data is spurious (Aitchison, 1986, 1997).

It is possible to formulate an hypothesis of independent compositions in the following terms. A composition has complete subcompositional independence if the subcompositions formed from any partition of the composition form an independent set. Hence, for additive log-normal compositions, the properties of complete subcompositional independence, and of being log-ratio uncorrelated, are essentially equivalent.

Subcompositional coherence

This vital concept can be best illustrated by a simple example. Consider two studies on deep-sea sediments. One investigator has dried his samples and then determined three oxides, A, B and C. A second investigator has samples from the same source, but has made his oxide determinations on the undried sediments and will, therefore, have results for not only A, B and C, but also H_2O. It will only take a brief moment to realize that product-moment correlations computed between the parts will be different for the two sets of analyses, but not the ratios between A, B and C. Hence, the product-moment correlation coefficient is subcompositionally incoherent for data involving parts, but as long as our statements are couched in terms of ratios, subcompositional coherence is maintained.

Aitchison (1997) gives a simple though telling illustration of what subcompositional coherence implies. Consider again the above example, this time backed up by figures provided by Aitchison:

Full compositions $(x(i), i = 1, \ldots, 4)$ *Subcompositions* $(s(i), i = 1, \ldots, 3)$

0.1	0.2	0.1	0.6		0.25	0.50	0.25
0.2	0.1	0.1	0.6		0.50	0.25	0.25
0.3	0.3	0.2	0.2		0.375	0.375	0.25

The product-moment correlation coefficient between the first two parts for the full set is 0.5. The scientist computing the correlation for the same two variables in the subcomposition would report a value of −1. There is thus **incoherence** of the product-moment correlation between raw components as a measure of dependence. However, if attention is directed towards the **ratio** of two components, it will be found that this remains unchanged for full compositions and subcompositions. Consequently,

$$s_i/s_j = x_i/x_j$$

because compositions expressed in terms of ratios are subcompositionally coherent.

Zero data values and compositional statistics

An uneasy aspect of compositional data analysis is the conceptual difficulty posed by zero observations and the taking of logarithms, an essential step in log-ratio multivariate analysis. Aitchison (1986) suggests various ways of circumventing this aspect. Here are some suggestions:

1 Check whether the problem can be solved by amalgamation. For example, it may be possible to proceed via the sum of alkaline oxides, rather than taking individual oxides separately. Consequently, if the content of Na_2O were zero, and that of K_2O were some small quantity in a few of a set of observational vectors, it might be deemed acceptable to pool sodium and potassium in the data set as "total alkalis".

2 Can an extremely small "mock-value" be entered? Some trace elements occur in close to negligible amounts such that in some samples, a "not-present result" is registered even though there may be minute quantities in the specimens. It is not an uncommon experience to find that trace elements previously registered as absent, come to light with the advent of vastly improved instrumentation. Developments in atomic absorption photo-spectrometry is a case that springs readily to mind.

3 Some very small quantity can be added to all the parts. This has the effect that all values are increased by the same minute amount and zero observations are made to disappear.

A word of warning is not misplaced here. If there are many "empty cells" in a data set it can be expected that a multivariate analysis, e.g. principal components, will exaggerate the contribution of the part concerned.

The log-ratio variance

The most useful solution to the multivariate analysis of compositions proposed so far is that of Aitchison (1986, 1997). The covariance structure of a D-part composition $\mathbf{x}$ is completely specified by the $\frac{1}{2}dD$ log-ratio variances, where $d = D - 1$:

$$\tau_{ij} = \text{var}\{\log(x_i/x_j)\} \quad (i = 1, \ldots, d;\ j = i+1, \ldots, D) \tag{1:4}$$

The corresponding means are:

$$\mu_{ij} = \mathbf{E}\{\log(x_i/x_j)\} \tag{1:5}$$

The inadvertency introduced by the log-ratio variance is, as just recorded, that the logarithm of zero does not exist, so if there are such observations in the data, and they are not uncommon in geochemical work, special procedures must be devised to deal with this situation. The most direct way of dealing with the question is by "preventative surgery", that is, to see whether the part giving rise to zeros is necessary to the project; if not, one could consider excluding it from the array before starting the analysis. We have all no doubt seen many a table of chemical analyses in which one component is always zero – it was analysed for, but found to be lacking in all samples. A second alternative is, as noted above, to add some minute number to all the entries in the crude data-matrix which is, however, not a good procedure. Such data are often referred to as *BDL* (which stands for below detection limit) observations. Other, more advanced, procedures are discussed in Chapter 11 of the book by Aitchison (1986).

THE LOG-RATIO DATA-MATRICES

It is frequently useful to have access to the log-ratio data matrices as a starting point. The program *logmat* performs the required calculations and stores the matrices for subsequent use.

Instructions for using the program **logmat**

Line 1: 1 the number of parts (number of columns)
         2 the size of the sample (number of rows)
         3 if the centred log-ratio data matrix is required, enter a 1 here. If the log-ratio covariance matrix is of interest, enter a 0. In this latter case there will be a second line containing the part to be used as a divisor, indicated by number. This, if the first part (entries in the first column of the data-matrix) is to be used as a divisor, a 1 is entered.

Thereafter follows the data-matrix in free format. The matrices are saved in the file *loggamma.dat* for the centred log-ratio data matrix and in *logsigma.dat* for the log-ratio data matrix. The trial data are in file *hongited.dat*.

MATRIX REPRESENTATIONS

The three representations now presented, the variation matrix, the log-ratio covariance matrix and the centred log-ratio covariance matrix may seem to be very different on first encounter, but they are in fact equivalent and each of them can be derived from either of the others by simple matrix operations (Aitchison, 1986, Chap. 4).

1 The variation matrix **T**

$$\mathbf{T}_{D \times D} = (\tau_{ij}) = [\text{var}\{\log(x_i/x_j); i, j = 1, \ldots, D\}] \tag{1:6}$$

This matrix is symmetric with a diagonal of zeros. Although it is not in the form of a covariance matrix it has certain computational advantages in that it treats the parts of a composition symmetrically (i.e. all parts are included on an equal footing). For many purposes in compositional data-analysis, the variation matrix is to be preferred.

2 The log-ratio covariance matrix Σ

$$\Sigma_{d \times d} = [\text{cov}\{\log(x_i/x_D), \log(x_j/x_D) \quad (i, j = 1, \ldots, d)\} \tag{1:7}$$

The log-ratio covariance matrix is the covariance matrix of a d-dimensional random vector $\mathbf{y} = \log(x_{-D}/x_D)$. (Note, that this vector is in space $\mathfrak{R}^d$.) Part D is held fixed, which means that the last component of the vectors of parts in the present representation, x_D, becomes the common divisor of all the log-ratios.

$$y_i = \log(x_i/x_D).$$

The complication posed by negative bias is eliminated by this transformation at the cost of a new difficulty in that one part is arbitrarily removed as a result of division by the common divisor x_D; this matrix does not, therefore, treat all parts of a composition symmetrically. Aitchison (1986, p. 92) proves that the order of parts, and the choice of a component divisor, does not influence the outcome of a multivariate analysis. The log-ratio covariance matrix is positive definite, which means, that it has a normal inverse.

3 Centred log-ratio covariance matrix Γ

A symmetric treatment of all D parts of a vector of compositions may be achieved by replacing the single component divisor x_D by the geometric mean of all D components.

For a D-part composition x, the $D \times D$ matrix

$$\Gamma = \text{cov}\left[\log\left(\frac{x_i}{g(x)}\right), \log\left(\frac{x_j}{g(x)}\right)\right] \quad i,j = 1, \ldots, D \tag{1:8}$$

is termed the **centred log-ratio covariance matrix**. This matrix is one that is useful for some multivariate analogues of full-space statistics. It is easy to interpret in that it has the advantage of being symmetric with respect to all parts. The drawback is that it is *singular* (which means that its determinant is zero, its rows sum to zero, and it does not possess a normal inverse) which places a particular (though not insurmountable) restriction on practical multivariate computational aspects.

The centred log-ratio covariance matrix (1:8) is the covariance matrix of the D-dimensional random vector:

$$\mathbf{z} = \log\{\mathbf{x}/g(\mathbf{x})\}$$

where $g(\mathbf{x})$ is the geometric mean of the parts, i.e.,

$$g(\mathbf{x}) = (x_1 \ldots x_D)^{1/D}$$

The singularity of this covariance matrix is due to the fact that $\mathbf{z}$ is confined to d-dimensional linear subspace and hence lacks the "freedom" of full space $\mathbb{R}^D$.

The most immediate obstacle in the path of many a multivariate analysis is how to obtain an inverse of the centred log-ratio covariance matrix (and correlation matrix). A generalized matrix inverse can be computed from the spectral relationship

$$\Gamma^- = \lambda_1^{-1} a_1 a_1^T + \cdots + \lambda_d^{-1} a_d a_d^T. \tag{1:9}$$

Equation (1:9) indicates that one computes the latent roots and vectors of Γ, then performs the reconstitution indicated for the reciprocals of the d latent roots that are greater than zero.

The three matrices just introduced are mutually interchangeble in the sense that one can pass from any of them to any other by a simple matrix manipulation (Aitchison, 1986, pp. 82–83).

SOME NOTES ON THE COMPUTATION OF SUBCOMPOSITIONS

From the point of view of manipulating subcompositions, the variation matrix **T**, defined in formula (1:6) on p. 23, is to be recommended because of its uncomplicated structure. It may not be immediately evident as to why an array of variances should be equivalent to the two covariance matrices introduced there, but the fact is that the specification of the variation matrix depends on variances of two-component log-ratios. Hence, the variation matrix appropriate to any subcomposition requires only the extraction from **T** of the elements in the rows and columns corresponding to the parts constituting the subcomposition.

If $\mathbf{T}_s$ denotes the variation matrix of the resulting subcomposition, **S** is the selecting matrix, of order $C \times D$, C is the dimensionality of the subcomposition, D is that of the full composition), then the matrix operation involved is:

$$\mathbf{T}_s = \mathbf{STS}^T.$$

The operations germane to subcompositions of the log-ratio covariance matrix and centred log-ratio covariance matrix are more complicated by far. In the latter case, not only the appropriate rows and columns of Γ must be selected but also the common log-ratio divisor must be changed from the geometric mean of all D compositions to that of the C selected parts. Construction of the log-ratio covariance matrix Σ_s is more intricate still, since all will depend on whether D is part of the subcomposition or not.

CONSTRUCTING A LOG-RATIO COVARIANCE MATRIX

The program *logcov* performs the steps required for producing log-ratio covariance matrices, namely, the log-ratio covariance matrix expressed by eqn. (1:7) and the centred log-ratio covariance matrix given by eqn. (1:8). The corresponding correlation matrices are also provided as well as the matching matrices for the crude data in order to permit comparisons to be made. It may be used as an alternative to the program *logmat*, introduced earlier on in this chapter.

The product-moment correlation coefficient has traditionally played an important part in geological analyses. However, in the case of the theory of compositional data, its relevance is much reduced owing to the deficiency known to statisticians as "incoherency". A reflection of this is that Aitchison (1986) hardly mentions it, confining his developments almost entirely to covariances. Our advice is to avoid thinking in terms of full-space correlations when dealing with compositions.

Instructions for using the program **logcov**

A set of trial data, encompassing five chemical parts, is located in the file *hongitln.dat*, a set of fictitious data constructed for illustrative purposes by Aitchison (1986). The input details for the observations to be run in the program are as follows:

    Line 1:   The number of samples to be treated.

Line 2:   The title of the job.
Line 3:   1 The number of specimens,

then,

2 the number of parts,

Line 4 and following: the listing of the data-matrix of parts.

If there is a second sample, the steps denoted above as "lines 2, 3 and 4" are repeated, but not line 1.

The input is in free format, which has the practical consequences that (a) you do not have to match up your data to fit some pre-ordained format, but (b), you must leave a space between each separate entry on the rows of the data-matrix.

For this first statistical computing example, it may be useful to explain how data-files are linked to a compiled statistical program, The recommended procedure is to type the following instructions on the command line. The *DOS* command is then

c:*program* < *infile* > *outfile*

where *infile* is the name of your data-set, arranged as explained above, and *outfile* is the name you wish to give to a file in which the results are saved. **Please note carefully that you cannot activate the program by clicking on the mouse – you must TYPE the instructions**. More explicitly, if the input file is called *cheman.dat* and the output is to go into a file you call *cheman.out*, then the appropriate command would be:

c:*logcov* < *cheman.dat* > *cheman.out*

You can then conveniently work on the file *cheman.out* by means of the editing facility EDIT of *DOS* 5 (and higher), or *Wordpad* in Windows 95 or *Graph Server*. Our advice is to you is to make full use of the Microsoft editing facilities for viewing the output of the set of programs and for preparing your own data for analysis. Both covariance matrices computed here are useful, as will become apparent as you encounter the multivariate techniques introduced further on.

The arrays for centred log-ratio correlations and log-ratio correlations and the corresponding raw correlations are listed in **BOX 1**: We have already stated that the correlation coefficient for compositional data lacks direct relevance and the material presented in **BOX 1** is provided merely for comparative reasons and not as a recommendation for use in interpreting comparisons with the discredited raw correlations of parts. A case can often be made in some work, however, for examining correlations between log-ratios of parts.

Box 1: Computing the log-ratio and centred log-ratio covariance matrices, corresponding correlations and comparison with the results using the procedure appropriate to full space. Data from Aitchison (1986). The log-ratio correlation matrices are shown for general information **and are not to be construed as an unconditional recommendation for general use**. The reasons for this statement are set out in the main text.

Program: **logcov**
Data: **hongitln.dat**

Hongite $N = 25$, a constructed data-set (Aitchison, 1986, p. 354)
Rows = 25 columns (parts) = 5

The log-ratio covariance matrix

	1	2	3	4
1	0.1386	0.2642	−0.2317	0.1214
2	0.2642	0.6490	−0.7132	0.1444
3	−0.2317	−0.7132	0.9627	0.0020
4	0.1214	0.1444	0.0020	0.1871

The centred log-ratio covariance matrix

	1	2	3	4	5
1	0.0661	0.1813	−0.2497	0.0164	−0.0140
2	0.1813	0.5557	−0.7416	0.0290	−0.0244
3	−0.2497	−0.7416	0.9993	−0.0485	0.0405
4	0.0164	0.0290	−0.0485	0.0496	−0.0465
5	−0.0140	−0.0244	0.0405	−0.0465	0.0445

The log-ratio correlation matrix

	1	2	3	4
1	1.0000	0.8809	−0.6344	0.7541
2	0.8809	1.0000	−0.9023	0.4144
3	−0.6344	−0.9023	1.0000	0.0047
4	0.7541	0.4144	0.0047	1.0000

The centred log-ratio correlation matrix

	1	2	3	4	5
1	1.0000	0.9460	−0.9719	0.2867	−0.2587
2	0.9460	1.0000	−0.9951	0.1748	−0.1553
3	−0.9719	−0.9951	1.0000	−0.2178	0.1921
4	0.2867	0.1748	−0.2178	1.0000	−0.9909
5	−0.2587	−0.1553	0.1921	−0.9909	1.0000

These two log-ratio correlation matrices are perforce different because they are correlating different log-ratios. However, their parent covariance matrices can be simply transformed from the one to the other. Note the two matrices differ in dimensionality.

Crude covariance matrix

1	29.7579	33.0209	−60.5583	−0.6607	−2.3299
2	33.0209	139.8515	−144.8595	−13.8257	−16.9022
3	−60.5583	−144.8595	185.0991	9.7125	15.9816
4	0.6607	−13.8257	9.7125	5.5458	−1.2630
5	−2.3299	−16.9022	15.9816	−1.2630	6.3579

Crude correlation matrix

1	1.0000	0.5118	−0.8159	−0.0514	−0.1693
2	0.5118	1.0000	−0.9003	−0.4964	−0.5668
3	−0.8159	−0.9003	1.0000	0.3031	0.4658
4	−0.0514	−0.4964	0.3031	1.0000	−0.2126
5	−0.1693	−0.5668	0.4658	−0.2126	1.0000

N.B. As we have already stressed, the "full-space" correlation counterparts are here shown solely for comparative and instructive reasons. We do not want this misconstrued as meaning that we want you to actually use them in your work.

The variation matrix and subcompositions

As observed on p. 24, the most efficient manner of dealing with subcompositional data is to proceed via the variation matrix. We shall now demonstrate this by means of two short programs, one for construction a variation matrix, **aitchvar**, the other for producing a subcompositional variation matrix, **subcomp**.

The trial-data for use with **aitchvar** are in **hongite1.dat**. The output consists of the appropriate variation matrix **T**. Save the result for later use in a file you may call **hongite.sbc**.

Instructions for using the program **aitchvar**

    Line 1:   1 sample size
                      2 number of parts

    Line 2+:   The data as an array of parts.

The resulting variation matrix provides the input to the next program, **subcomp**, which computes the product

$$\mathbf{T}_s = \mathbf{STS}^T$$

The program also provides Aitchison's (1997) finite scale transformation of the variation matrix (which goes part of the way in replacing the correlation matrix of full space).

Input details for subcomp:

    Line 1:   1 Number of parts in the full composition (D)
                      2 Number of parts in the subcomposition (C)

    Lines 2+: The full variation matrix **T**.

    Lines 3+: The selecting matrix **S**.

The selecting matrix (p. 26) requires care in its formulation. It is of order $C \times D$, that is, it has C rows and D columns. Each row contains a single element equal to one, and at most one in each column. Examine the display in file **hongite.sbc** with the output just obtained. There has been a reduction from the five-part variation matrix to the one appropriate for the subcomposition (x_1, x_4, x_5). A little experimenting with the selection matrix will make you familiar with the use of the program. Note, that the compositional variance matrix **always has zero diagonal elements**.

Checking a data-matrix for constant sum

Usually, you will be able to see by simple inspection whether or not a data-matrix is constrained. In cases where this is not obvious (very large data-matrices, for example) you can check for "closure" by running the program *propmat*. This program returns row-sums and column-sums. If either of these yields a constant sum, row-by-row, or column-by-column (the data-set can be in transposed form), then you have constrained data of the most common type. The program will not, of course, unveil the constraint imposed on observations made on samples of "constant weight", nor data-matrices of frequencies of objects produced by counts to a constant total.

Instructions for using the program **Propmat**

The input specifications for running the program are:

Line 1: the number of variables/parts (=columns) (comma) then the number of specimens (=rows). In the trial example, called *boxitprp.dat*, this instruction reads

5,10

Line 2, and following: the data-matrix.

Sometimes the row-sums will not be exactly the same. There are several valid reasons for this. Chemical determinations often contain an error component, which can cause slight deviations from 100%. (This topic is well covered in Aitchison's book.) Also, some minor component, or components, may have been omitted from a published table of values. An example is the file *haiticvt.dat* in which some minor parts have been deleted. Sometimes the analyst may have subtracted values from his data in the widely held though completely mistaken belief that this will "open up the constraint" (frequently referred to dramatically as "breaking open the data" or "shattering the constraint"). The constraint is, however, incorporated into the covariance structure and, hence, it is not possible to remove its effects just by leaving out one or more of the variables.

To use the program, type *propmat* at the *DOS* prompt. You will be asked to supply the name of the file containing the data-matrix to be examined. The output shows each row-sum in turn then the column-sums. The data for the fictitious rock-type "boxite" sum nicely to 100, as they were made to do by their inventor.

An application to a constant sum matrix

It can be instructive to look at the output yielded by the program *propmat* applied to a real situation. We have chosen some data from the Ocean Drilling Project series of volumes. Usui (1992) studied hydrothermal manganese minerals in cores from Leg 126 of the Ocean Drilling Programme in the Japanese region. Among the published data, we selected the analyses listed in his Table 2 to exemplify the problem of analysing compositional data. The compositions of manganese minerals included in the project by Usui are the oxides of Mn, Fe, Na, K, Ca, Mg, Ba, Si and Al, being nine in all. Program *propmat* shows that the rows do not sum to 100% (the quantities are cited in weight percent) and something more than water (said by Usui to be around 7% lost on heating) must have been excluded from the published table. This example is enlightening because it illustrates the dangers lurking neath an inappropriate statistical application. This was, alas, not realized by Usui, who based his interpretations on the usual (=crude) correlation coefficient between parts.

Geological compositional data tend to deviate from the multivariate Gaussian state with respect to kurtosis and skewness and more extreme divergencies from theory can influence stability in the compositional covariances; it is therefore advisable to carry out the appropriate tests of normality as a first step to a serious analysis and to check the data for atypicalities. This topic is further considered in a later section.

Compositional analogue of the correlation coefficient

For "normal" (=full-space) data, there can be no confusion in the choice of the appropriate product-moment correlation coefficient – there is only one available. Matters are less simple for compositional data in that there are two ways of expressing a matrix of covariances, but neither of these leads to a sensible correlation matrix in product-moment terms. Aitchison (1986, p. 105) considered this situation from the viewpoint of validity of log-ratio-uncorrelated compositions and provided a set of criteria for general guidance.

A prime preoccupation among petrologists/geochemists is the desire to express relationships between pairs of elements or oxides. In many publications, the only statistical steps taken are the computation of product-moment correlations. Aitchison (1986, 1997) has demonstrated that the concept of zero correlation has no meaning in simplex space. The question arises then as to whether there is some possible analogue to the concept of correlation for compositional data. Let us summarize some of the relevant properties of compositions:

1 For compositions, the sizes of the specimens are irrelevant. The composition of a rock is dimensionless and scale-invariant.

2 Any meaningful function of a composition can be expressed in terms of ratios of the components of the composition.

3 Compositions are concerned with *relative values* – therefore with ratios of components, and not with absolute magnitudes.

4 The property of subcompositional coherence. Subcompositional coherence is lacking for product-moment correlations of raw components.

5 The Variation Matrix, **T**, (cf. eqn. 1:6) is an appropriate summarizing measure of dispersion and dependence. It cannot, however, be expressed as a conventional covariance matrix of some vector (although Σ, Γ, and **T** are equivalent). The nearest one can come to the product-moment correlation is the *relative variance*,

var $\{\log(x_i/x_j)\}$. Thus, for example, $\tau_{ij} = 0$ means that there is a perfect relationship between x_i and x_j in the sense that their ratio is constant (remember, the diagonal elements of the variation matrix are all zero). In other words, the concept of perfect correlation between variables is replaced by that of *perfect proportionality* between parts. The greater the value of τ_{ij}, the greater is the departure from perfect proportionality: moreover, as this value approaches infinity, the condition of a complete lack of proportionality is neared.

To make the idea of proportionality more accessible to the average practitioner, Aitchison (1997) introduced a finite scaling transformation, $\exp(-\sqrt{\tau_{ij}})$, as a measure of the relationship between two parts. This scale runs from 0, which signifies a lack of proportional relationship, to 1, that corresponds to a perfect proportional relationship. (N.B. It requires at least three parts for the computation of the proportionality measure.) A simple example appears in **Box 2**. This transformation is not in any way meant to be the compositional counterpart of the full-space correlation coefficient. It does, however, provide a way of expressing strength of relationship between two parts though, unfortunately, without identification of a negative or positive association.

Additionally, more direct hypotheses of independence can be formulated in terms of the independence of subcompositions (Aitchison, 1997, eqn. 19).

Box 2: Interpretation of the Variance Matrix: the hongite constructed data-set. There are 5 "oxides" and 25 specimens.

The variance matrix

	1	2	3	4	5
1	0.0000	0.2593	1.5647	0.0828	0.1386
2	0.2593	0.0000	3.0381	0.5473	0.6490
3	1.5647	3.0381	0.0000	1.1458	0.9627
4	0.0828	0.5473	1.1458	0.0000	0.1871
5	0.1386	0.6490	0.9627	0.1870	0.0000

The lowest degree of proportionality occurs for combinations with part 3 and the highest for combinations with part 4.

Aitchison's finite scale transformation

	1	2	3	4	5
1	1.0000	0.6010	0.2863	0.7499	0.6892
2	0.6010	1.0000	0.1750	0.4772	0.4468
3	0.2863	0.1750	1.0000	0.3429	0.3749
4	0.7499	0.4772	0.3429	1.0000	0.6489
5	0.6892	0.4468	0.3749	0.6489	1.0000

The finite scale transformation can almost be claimed to emulate a correlation profile. Be not deceived, however, the interpretation rests on different grounds. We can see, however, that the highest level of proportionality is for the pairing part 1/part 4 and the lowest for the pairing part 2 with part 3.

AN APPROXIMATION FOR COMPOSITIONAL VARIATION ARRAYS

Can anything be done to make the host of inappropriate analyses of petrological and geochemical data accessible for modern analytical procedures? There is a possibility, providing that the published material contains a fully reported set of results. With access to the crude covariances (or crude correlations and crude standard deviations) and the crude means, a good to fairly good approximation to the log-ratio means and covariances can be obtained by applying the theory of standard approximations for means and covariances of functions of random variables (Cramér, 1946, Aitchison, 1986). The program *appr* performs the requisite calculations. The trial-data are in the file *boxapp.dat*.

Instructions for using the program **appr**

Input

Line 1: The dimensionality of the input matrix of raw covariances.

Lines 2+: The raw covariance matrix.

Last line: The means.

Output

A matrix containing the compositional means in the lower triangle and the compositional variances in the upper triangle. A specimen of the output is given below. It is for the boxite data of Aitchison (1986). The input data was obtained by performing a standard (inappropriate) computation of covariances and means on the raw data.

Approximate log-ratio means and variances computed from raw variances and covariances and the raw means for the boxite data.

lower triangle = means, upper triangle = variances

```
    1       2       3       4       5
 0.0000  0.1360  0.0555  0.0875  0.0319
 0.4055  0.0000  0.1958  0.1896  0.1310
 0.9931  0.5876  0.0000  0.1085  0.0542
 1.1573  0.7518  0.1642  0.0000  0.0588
 1.7011  1.2956  0.7080  0.5438  0.0000
```

THE BOX–COX TRANSFORMATION

Barceló et al. (1996) have raised some important points concerning the transformation of compositional data and the identification of "outliers". The basis of this contribution lies with results of Aitchison (1986), but because of the theoretical unattractiveness of the methods evoked, Aitchison did not make much use of them in his book. One of these, the Box–Cox transformation, has the advantage that it can often provide a better fit than a logarithmic transformation, but suffers from theoretical complications and the fact that it is greatly affected by the divisor chosen for computing it. The second method considered by Barceló et al. (1996) is the additive log-ratio transformation (Aitchison, 1986, p. 113), which is permutation invariant (the choice of divisor does not disrupt multivariate normality). The need for the robust estimation of multivariate parameters is underlined which is a subject that has been taken up by many authors, including Gnanadesikan (1977), Campbell (1980), Campbell and Reyment (1980), Gordon (1981), Hadi (1992) and Hampel et al. (1986).

The identification of types of "outliers" has been given careful consideration by Krzanowski (1987a, 1987b). For the purposes of our primer, we have found this approach that uses a jackknife technique, useful for many practical purposes. For a recent appraisal of "bootstrapping" and "jackknifing", see Efron (1992). Probability plotting offers a useful means of making a preliminary appraisal of the normality status of a sample, but it cannot supplant an analytical evaluation of the properties of a data-set.

MULTIVARIATE NORMALITY

Mardia (1970) derived large-sample tests for deviations from multivariate skewness and kurtosis, which were applied in geological connections by Reyment (1971, 1991). Briefly, the measures of skewness and kurtosis were defined by Mardia as follows. If

$$g_{rs} = (\mathbf{x}_r - \bar{\mathbf{x}})\mathbf{S}^{-1}(\mathbf{x}_s - \bar{\mathbf{x}})$$

the measure of multivariate skewness is

$$b_{1,p} = \frac{1}{n^2} \sum_{r,s=1}^{n} g_{rs}^3. \qquad (1:10)$$

The measure of multivariate kurtosis is

$$b_{2,p} = \frac{1}{n} \sum_{r=1}^{n} g_{rr}^2. \qquad (1:11)$$

These moments are invariant under *affine* transformations. $b_{2,p}$ picks up extreme behaviour in the Mahalanobis generalized distances of objects from the sample centroids. Note, that these are large sample tests and cannot really give a decisive answer for sample-sizes of less than 50 specimens (Seber, 1984). A simple example will suffice to demonstrate the capabilities of the program *multnorm.exe*. The trial data are in *ivuna.dat* and consist of chemical analyses on carbonates in CI chondrites (Endress and Bischoff, 1996). The study of chondrites forms part of the field of meteorite research. All CI-chondrites are regolith breccias consisting of various types of chemically and mineralogically distinct mineral and lithic fragments. The chemical components determined on a chondrite from Ivuna are CaO, MgO, MnO, FeO, and CO_2. The two authors were concerned with correlation problems. Apart from the fact that they did not use the appropriate form of the correlation coefficient, inspection of the tables suggests that there could be marked non-normality in the data. This was borne out by the analysis for multivariate normality, the results of which are summarized below.

multivariate skewness $b_{1,p} = 40.168$

multivariate kurtosis $b_{2,p} = 61.928$

Significance results

Multivariate skewness chi-square $= 234.31$ for 35 degrees of freedom.

Significance level for chi-square (95% level) $= 49.52$

BETA for multivariate kurtosis is $= 9.52$

Summary

Significance may be assessed from a table of the standard normal variate (available in all books of statistical tables, for example, Rohlf and Sokal, 1969). Clearly, the data deviate markedly from multiviariate normality both with respect to skewness as to kurtosis. With this result in hand, the analyst would be well advised to probe further the properties of his data.

Instructions for using the program **multnorm**

Input

Line 1:   contains the number of variables, then the number of observations

Line 2:   the data-matrix

Output

Intermediate steps are listed, including the sample mean vector, the covariance matrix and its inverse. The skewness and kurtosis statistics follow, together with the results of the tests for significance. The trial-data are in *ivuna.dat*.

SUMMARY

1 Definitions of statistical entities
2 Basic matrix operations.
3 Compositional data and the log-ratio transformation.
4 Basic matrices for compositions
5 Multivariate normality
6 Programs and associated training sets

covmat	haiticvt.dat
matops	matops1.dat
matinv	matinv1.dat
logmat	hongited.dat
logcov	hongitln.dat
aitchvar	hongite1.dat
subcomp	hongite.sbc
propmat	boxitprp.dat
appr	boxapp.dat
multnorm	ivuna.dat

Chapter 2

Graph Server, the GS Language and Graph Wizard

INTRODUCTION

This chapter describes the use of graphic software available on the CD-ROM accompanying this book. An on-line version of this documentation is available on the CD-ROM in the form of HTML files. These files can be viewed by using a web browser. If you have a web browser correctly installed on your machine, double-clicking on an HTML file in Windows Explorer will launch the browser and load the file. The table of contents of the HTML documentation is located in the file *Graph Server**Documentation**default.html* on the CD.

The documentation on the CD-ROM contains information on how to install *Graph Server* and *Graph Wizard* (read the file *readme.txt* or the HTML documentation). The documentation on the CD-ROM contains also updated information on the use of these programs.

This chapter is organized into six sections:

Section 1. Using Graph Server. Graph Server is a program which interprets a sequence of graphic commands and generates a corresponding display. Generated displays can be saved as files or printed.
Section 2. The GS Language. GS is an interpreted language built into Graph Server. A library of functions built into GS provides low-level graphic primitives and a few higher-level functions useful when drawing graphs. This section describes the syntax and scope of the GS language and contains a complete reference of the GS functions.
Section 3. Debugging facilities of Graph Server. Graph Server allows you to inspect the GS function calls and symbolic constants it received from client programs. These facilities can be used to debug GS code.
Section 4. A GS Tutorial. Provides several examples of how to write GS code.

Section 5. Graph Wizard. Graph Wizard guides the user through a series of interactive steps to generate a scatter plot from a data file. Once the procedure is complete, Graph Wizard generates the corresponding GS code and sends it to Graph Server.

Section 6. Frequently Asked Questions. This section concentrates on questions likely to arise during the practical use of Graph Server.

SECTION 1. USING GRAPH SERVER

Software requirements

Graph Server runs under Windows 95 and Windows NT (Workstation or Server) version 4.0 or later. There is no functional difference in Graph Server when running under Windows 95, Windows 98 and Windows NT, but graphs generated under Windows 95/98 may show small differences in details (e.g., in the way vertices of zigzag lines are drawn) with respect to the same sequence of GS commands issued under Windows NT. This is due to the fact that there are small differences in the graphic libraries built into either operating system. As a whole, Windows NT version 4.0 behaves more predictably than Windows 95/98.

Graph Server does not run under DOS, Windows 3.1, Windows for Workgroups 3.11, Win32s and versions of Windows and Windows NT earlier than those mentioned above.

There are no versions for other operating systems.

Hardware requirements

Graph Server runs on all Intel and Intel-compatible processors of the x86 family that are supported by Windows 95/98 and Windows NT 4.0. Graph Server has no built-in limitations with respect to the CPU type, but machines with an Intel Pentium II, Celeron or higher processors are recommended because of speed considerations. At least 64 MB of RAM are recommended, and larger amounts may yield faster processing. Each graph displayed on the screen is contained in a separate instance of Graph.exe, which uses a few MB of memory. Therefore, on a system with limited memory you should close all graph windows once they are no longer necessary.

There are no versions for other CPU types.

Installation of Graph Server

Graph Server consists of the files *Graph.exe*, *GraphClient.dll* and *GraphClient.ini*. These three files should be copied to your Windows directory. On most Windows 95 and Windows 98 machines, this is *C:\Windows*. On most Windows NT machines, the Windows directory is *C:\Winnt*. You should also edit GraphClient.ini so that the *Graph =* statement points to the complete path to the location of Graph.exe. For instance, on most Windows 95/98 machines, this line should read:

Graph = C:\Windows\Graph.exe

Note that GraphClient.ini contains alternative Graph= statements for typical Windows 95/98 and Windows NT, but only one such statement should be active. Inactive statements begin with // (two consecutive forward slashes). Note also that the *[Settings]* heading should be located at the beginning of the file.

Graph Wizard is a program available on the book CD, but is not, strictly speaking, a part of Graph Server. Instead, it is an application program that uses Graph Server to generate graphs (see below). You can copy *GraphWizard.exe* from the CD to any location you like (a subdirectory of *C:\Program Files* or *C:\Programs* is most suitable). Subsequently, from Windows Explorer you can create a shortcut to Graph Wizard. One way of doing this is:

- Right-click on GraphWizard.exe in its new location, and select *Create Shortcut* from the menu. This creates a shortcut, named *Shortcut to GraphWizard.exe*, in the same directory.
- Drag the shortcut to a suitable location, e.g. the desktop.
- Left-click on the shortcut name twice (do not double-click, just click twice with a short pause between clicks). This allows you to edit the shortcut name.
- Write a new name for the shortcut (*Graph Wizard* is most appropriate).
- Remove the focus from the shortcut name by clicking something else.

Now you can double-click on the shortcut whenever you want to start Graph Wizard.

Scope of Graph Server

At the outset of this project, we decided to add modern graphic capabilities to a set of programs that were originally written as research tools. Although scientifically sound, and efficient from a computational point-of-view, these programs were not designed to run under the Microsoft Windows family of operating systems. In fact, most of these programs lacked any graphic capability. Since the scope of this book is to enable scientists to use the mathematical methods discussed in the text, and since using these methods often means, in practice, incorporating them into one's own program, it was not feasible to re-write the programs for the Windows environment.

The principal problem with the above approach (in addition to the effort necessary to convert all these programs) is that Windows programs are event-driven. Their source code lacks the structure most non-professional programmers are familiar with. For instance, a Windows program written in C does not have a *main( )* entry-point. Functions relating to the graphic or user interface are not called by other parts of the program. Instead, different portions of the program communicate with each other, and with the user, by retrieving and replying to messages at the request of the operating system. The operating system stores messages in queues and dispatches them to the proper software units.

Since readers cannot be expected to learn to program for the Windows environment, it was decided to follow a different approach. It was necessary to isolate the programmers from the drudgery of the Windows internal workings. Although it was not possible to create a complete abstraction from the design of the operating system, a measure of device-independence could be achieved. Instead of communicating directly with Windows, a program can be written to issue a series of graphic commands to a server, which in turn displays their result in graphic format. Thus, even a DOS-mode program can produce graphic output in a separate display window. Graph Server is designed to achieve this goal.

Graph Server is not a stand-alone graphic program that works interactively with the user. Instead, input to Graph Server is provided by another program that generates numeric data and issues graphic commands. Therefore, users cannot directly "edit" a chart on-screen. Instead, they can modify and re-run the program that issues the graphic commands. Editing of a graph, in the sense of cropping it, masking portions of it, changing colours and adding labels, can be done as a post-processing stage, by saving the graph and loading it into a graphic editor.

Several commercial programs provide many of the charting facilities needed by scientists. These programs are either stand-alone products, or add-ons to generic spreadsheet packages. We felt that it was unnecessary to duplicate the capabilities of these programs, and instead decided to leave the user in complete control of the graph being generated. In practice, this means that the user is responsible for the placement of each element composing a graph from the reference axes and ticks to the data points. This involves more work on the part of a programmer. On the other hand, the end-result is limited only by the display capabilities intrinsic in Windows and in the graphic hardware.

Because of its inherent flexibility, Graph Server is not limited to graphs, and can be used as a graphic-display module for programs that are not written for the Windows programming interface.

Architecture of Graph Server

Graph Server is implemented as a client-server system. The charting program *sensu stricto* is a server that receives commands from a client program (i.e., a program written by the user), parses and processes them, builds a graphic representation

Section 1. Using graph server 43

of the chart, and displays it (printing it or saving it as a file upon request). Several client programs may use Graph Server simultaneously. Each client controls a separate instance of Graph Server.

There are two ways to use Graph Server. The most comfortable way is to make your program use GraphClient.dll, which is a file containing all the code necessary to communicate with Graph Server. An alternative is to make your program save all graphic commands to a text file. Subsequently, this file can be sent to Graph Server directly (see below) or by the *SendFile.exe* utility. This causes graphic commands to be routed directly to the server. The second alternative may be convenient at an early stage in the development of your program, or if your compiler does not support the interfacing to dynamic-link library (DLL) files.

Graph Server is a state machine: the effects of commands are stored internally as settings that remain valid until changed by a subsequent command. In practice, subsequent processing by the server continues to overlay graphic elements to the image, until a command to clear the image is issued. The server also stores internally the complete sequence of commands issued by the client during a session. Clearing an image causes the stored sequence to be erased.

Running Graph Server

You don't need to start Graph Server before using it. It will start automatically whenever a program will call the functions contained in GraphClient.dll (see below). In the present version, you also can start Graph.exe from a DOS box or from the *Start → Run* menu and provide a file name on the same command line. Graph.exe will read the file (which should contain GS statements; see below) and display the corresponding graph. In addition to the file name, you can specify the **−d** switch to delete the input file after reading it, and/or the **−l** switch to create a log file. These features may change in future releases.

Introduction to the use of DLLs (Dynamic Link Libraries)

A DLL file is a compiled library of functions that are linked to a program at run-time (as opposed to static linking, which permanently embeds a library in the program's executable file). The main advantages of using a DLL are:

- Several programs can simultaneously share the code in a DLL, while only one copy of the DLL needs to be present.
- The DLL code is not embedded in the executable file, which reduces its size.
- Upgrading a DLL does not require recompilation of the programs that use it. Only the DLL file needs to be changed.

The trade-off of using a DLL is that calling its functions is slightly more complicated than calling those in a statically-linked library. Examples of this procedure, which is language- and compiler-specific, are given in the next section.

Writing a program that uses GraphClient.dll

Client programs communicate with Graph Server by using code in a dynamic-link library located in the file GraphClient.dll. This file is included in the software accompanying this book. You should copy the file to one of the following locations:

- The Windows directory (typically C:\WINDOWS in Windows 95, or C:\WINNT in Windows NT).
- The system directory (typically C:\WINDOWS\SYSTEM or C:\WINNT\SYSTEM32). This is the preferred location for DLL files.
- The directory containing your executable file that links to GraphClient.dll.
- A directory specified in the PATH environment variable.

There is no functional disadvantage in having two or more copies of a DLL file in the locations listed above. However, a reason for not doing so is that, should the DLL be upgraded to a later version, some of the older copies may be left over, with unpredictable results. DLLs placed in a directory other than those listed above cannot be found by the operating system.

Programming interface

GraphClient.dll contains three functions that can be called by the client program. Their prototypes (as contained in the source code of the library) are:

extern "C"_declspec(dllexport) BOOL Connect();
extern "C"_declspec(dllexport) const char* Communicate(char* pData);
extern "C"_declspec(dllexport) void Disconnect();

The non-ANSI keywords and macros that precede each function declaration tell the compiler that the functions must be made accessible to external programs using the DLL. Since the DLL is written in C++, the empty argument lists in the **Connect()** and **Disconnect()** prototypes imply that the functions take no argument (as opposed to *any* argument in C). The type **BOOL** is defined (in the include file *windows.h*) as an **unsigned int**. A **BOOL** can take the values **TRUE** (non-zero) or **FALSE** (zero). Removing the compiler- and system-specific parts, and converting to C, yields prototypes more familiar to C programmers:

unsigned int Connect(void);
const char* Communicate(const char* pData);

Section 1. Using graph server

void Disconnect(void);

Connect() must be called when the client program wishes to connect to Graph Server. The function returns a non-zero value if it succeeds, zero otherwise. Your program should check this value before proceeding, and act accordingly. A failure to connect means that no further command can be issued to the server (except for a new attempt to connect).

Communicate() is called to send a string containing a GS command to Graph Server. This function takes as argument a pointer to a null-terminated string. Commands must follow the syntax of the GS language (see section 2 of this chapter). If the command or statement is successfully sent to the server and the server replies, the function returns a pointer to a null-terminated string containing the server's reply. This function returns a pointer to the string "OK" if the command is sent to Graph Server successfully (this only means that Graph Server received it, not necessarily that it parsed and executed it without errors). If something prevents communication, the function will return a null pointer. Therefore, your program should check that the pointer is not null before attempting to read the server's reply. Attempting to use a null pointer will cause your program to crash.

A communication session with Graph Server typically consists of a large number of calls to **Communicate()**.

Disconnect() should be called once your program has finished sending commands to the server. You cannot call **Communicate()** after calling **Disconnect()**. Note that calling **Connect()** after disconnecting will *not* resume the previous session. Instead, it will create a new instance of the server program and draw the new graph in another window. There is no built-in mechanism to allow a single client to use two or more *simultaneous* instances of the server.

The recommended (but not the only) way of calling the DLL functions from Microsoft C and C++ is shown below. Depending on the language and compiler you are using, there may be other ways. This is an example of a C program that connects to the server, issues a single instruction and disconnects.

```
#include <windows.h> //contains function and type definitions
// necessary to use DLLs
void main(void)
{
    HINSTANCE hDll; // you can regard this as a handle to the DLL
    // we declare pointers to the DLL functions
    BOOL (FAR* pConnect) ();
    const char* (FAR* pCommunicate) (char* pData);
    void (FAR* pDisconnect) ();
    // this variable will contain the return value of Connect()
    BOOL bRet;
    // this variable will contain the server's reply
```

```
            const char* pReply;
            // load the DLL into memory
            if(hDll = LoadLibrary("GRAPHCLIENT.DLL"))
            {
                // get pointers to the DLL functions
                pConnect = (BOOL(FAR*)()) GetProcAddress(hDll, "Connect");
                pCommunicate = (const char*(FAR*)(char*)) GetProcAddress(hDll,
                "Communicate");
                pDisconnect = (void(FAR*)()) GetProcAddress(hDll, "Disconnect");
                // call the DLL functions (see discussion below)
                if(bRet = (*pConnect) ())
                {
                    pReply = (*pCommunicate) ("// This is a command. ");
                    // Your program should check whether
                    // Graph Server replies with "OK".
                    (*pDisconnect) ();
                }
            }
        }
```

The code used to declare and call the DLL functions may look baffling to C programmers who have not yet mastered pointers (which is usually regarded as the Great Divide between the beginner and experienced levels). However, the apparent complexity of the code is easily explained. The principal point to note is that the DLL functions are declared as pointers to functions, *not* as function prototypes. In simplified terms, a pointer to a function can be thought of as the address in memory where the function's code is stored. The declaration of a pointer to a function must enclose the indirection operator * and the function name within parentheses, lest the compiler applies the indirection to the function return type.

These pointers are subsequently initialized by calling **GetProcAddress()**, which returns the values of pointers to the corresponding DLL functions. The casts to a function pointer with specific return and argument types are necessary to convert the return value of **GetProcAddress()**, which is a generic function pointer.

In order to call the functions, the pointers must be indirected by using the operator *. The notation used in the function calls, therefore, parallels the notation used to indirect pointers to variables:

```
char* p;
*p = "Hello";
```

where **p** is a pointer to a **char**, and *p is a **char** variable. Similarly, **pConnect** is a pointer to a function, and *pConnect can be viewed as the corresponding function. Lastly, the indirected function pointer must be enclosed within parentheses, because

Section 1. Using graph server

the parentheses enclosing the arguments of a function have a higher associativity than the indirection operator. As with pointers to variables, alternative notations can be used to declare and use pointers to functions, but none of them offers substantial advantages over the others.

If you feel you do not completely understand the above discussion, try to write your code exactly as shown in the example. It should work without problems.

A slightly simplified version of the source code of SendFile.exe, a utility that uses GraphClient.dll, is shown below as an example.

Source code of SendFile.exe

```
// SendFile.c

#include <windows.h>
#include <stdio.h>

void main(int argc, char** argv)
{
    HINSTANCE hDll;    // handle to DLL
    BOOL (FAR* pConnect) ();   // DLL function prototypes
    const char* (FAR* pCommunicate) (char* pData);
    void (FAR* pDisconnect) ();
    BOOL bRet;
    char pMsg[32000];    // buffer for message
    const char* reply;
    int i;
    FILE* pFile;

    // load the dll
    if(!(hDll = LoadLibrary("GraphClient.dll")))
    {
       puts("Cannot load GraphClient.dll");
       return;
    }

    // get the pointers to the dll functions
    pConnect = (BOOL(FAR*)()) GetProcAddress(hDll, "Connect");
    pCommunicate = (const char*(FAR*)(char*)) GetProcAddress(hDll, "Communicate");
    pDisconnect = (void(FAR*)())GetProcAddress(hDll, "Disconnect");

    // call the dll functions
```

```
if(!(bRet = (*pConnect)()))
{
   puts("Cannot initialise Graph Server");
   return;
}

pFile = fopen(argv[1], "r");
if(!pFile)
{
   puts("Cannot find or open the input file.");
   return;
}

do
{
   i = -1;
   do
   {
      i++;
      if(!fread(pMsg + i, 1, 1, pFile))
      {
         *(pMsg + i) = '\0';
         break;
      }
   } while(pMsg[i] != '\n' && pMsg[i] != '\0');
   *(pMsg + i) = '\0';
   if(!strlen(pMsg))
      break;
   puts(pMsg);
   reply = (*pCommunicate)(pMsg);
   if(strcmp(reply, "OK"))
   {
      puts("Error while communicating with GraphServer.");
      goto cleanup;
   }
}
while(TRUE);

cleanup:
   (*pDisconnect)();
}
```

Section 1. Using graph server

A note is necessary to explain why direct access to the internal operation of GraphClient.dll and Graph.exe is not provided. These programs contain the routines to implement the actual communication between client and server. The communication mechanisms have changed radically several times during the development stage of Graph Server, and may change again in future releases. Since these mechanisms are completely isolated from users, user-written programs that call the interface functions in GraphClient.dll will need no change in order to run with different versions of Graph Server. The only necessary change will be the installation of the new Graph Server files.

Writing a program that uses SendFile.exe

In case you should find it impossible or impractical to use GraphClient.dll with code generated by your compiler, you can still use Graph Server. As a first step, your program must write the GS instructions to a text file. Each instruction should be placed in a separate line. Then, run the program SendFile.exe (included with the software accompanying this book), specifying the name of the text file on the command line. If the name of the text file contains spaces, enclose it within quotation marks. For example:

SENDFILE "C:\data\my file.txt"

SendFile.exe reads one GS instruction at a time from the file, displays it on-screen and sends it to the server.

SendFile.exe is provided principally as a stopgap measure. Whenever possible, you should use GraphClient.dll from within your program, because in this way your program communicates directly with the server. Incidentally, SendFile.exe uses GraphClient.dll to communicate with GraphServer.

Viewing a graph

Once Graph Server has received a complete set of GS statements and the client has disconnected, the graph is displayed. Initially, the graph fills the display window. You can use the View → Zoom In and View → Zoom Out menu items (or the corresponding buttons on the toolbar) to inspect details in the graph. You can use the View → Zoom to Fit menu item to return to a full view of the graph. You may use this button also if you lose track of where in the graph you are scrolling, and want to return to a full view.

In order to scroll a graph, place the mouse cursor on the window surface. The cursor will change to an open hand. Press and hold down the left mouse button. The hand will close. Drag the cursor in the direction you want to scroll to. The graph will scroll once you release the mouse button. This mechanism is especially effective if you place the cursor on top of a graph element, press the left button and drag the mouse to a new location where you want the element (and the whole graph surrounding it) to be moved.

It is possible to scroll a graph also when it fits entirely within the window. This may be useful if you suspect that you have misplaced a graph element, so that it lies somewhere in the area outside the graph frame.

Each graph is contained in a separate instance of graph.exe, i.e., a self-standing program that runs in its own memory space. Therefore, depending on the amount of RAM available in your computer, there may be a practical limit to the number of graph windows that can be left open simultaneously.

Printing a graph

Graph Server allows you to print a graph. This command is available under the **File → Print** menu item.

Setting the page margins

Before you print a graph, you may wish to set its size. This is done in the **Page Setup** dialog box, which is displayed when you select the **File → Print** menu item. This dialog box (below) displays a small preview of the page margins. Note that the graph displayed in this preview is not your graph, but just a standard icon. The only purpose of this preview is to give you a visual impression of the size of the margins.

You can enter the numerical value of the size of the margins in the appropriate controls. Your graph will be made to fit the available print area (i.e., the page area within the margins). Note that the aspect ratio (i.e., the ratio of height to width) or your graph will not be altered. In other words, the graph will not become vertically or horizontally "squashed".

Printing the graph

Once you are satisfied with the margin settings, you can press the **OK** button in the **Page Preview** dialog box. This displays the **Print** dialog box (below), in which you can set the number of copies and access other printer settings. As a rule, the availability of these settings depends on the type of printer you are using.

Once you press the **OK** button, Graph Server will send the current graph to the printer.

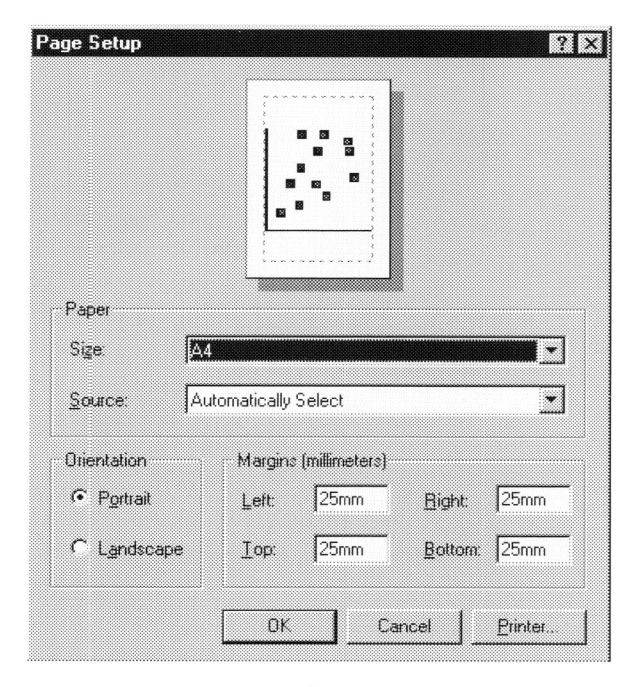

Fig. 1

Fig. 2

Previewing the printout

You can use the **File → Print Preview** menu item to examine the appearance of your graph as it will appear once printed on paper.

The preview image may differ in some details from the image displayed in Graph Server's document window. For instance, if your printer cannot print in colour, text will be displayed in black, regardless of its original colour in the graph. Other graphic elements, instead, will retain their colour in the preview window, even though the printer may not be able to print in colour. These differences reflect the standard characteristics of print previews as implemented in all Windows applications.

Saving a graph

Graph Server allows you to save a graph to a file, and to choose between two types of file. The first type is a bitmap, i.e., a rectangular matrix of pixels. The second type is a metafile, and consists of a sequence of graphic commands. Unlike the GS code accepted as input by Graph Server, the Windows Enhanced MetaFile format used to save a graph is a standard understood by several graphic editors (see below).

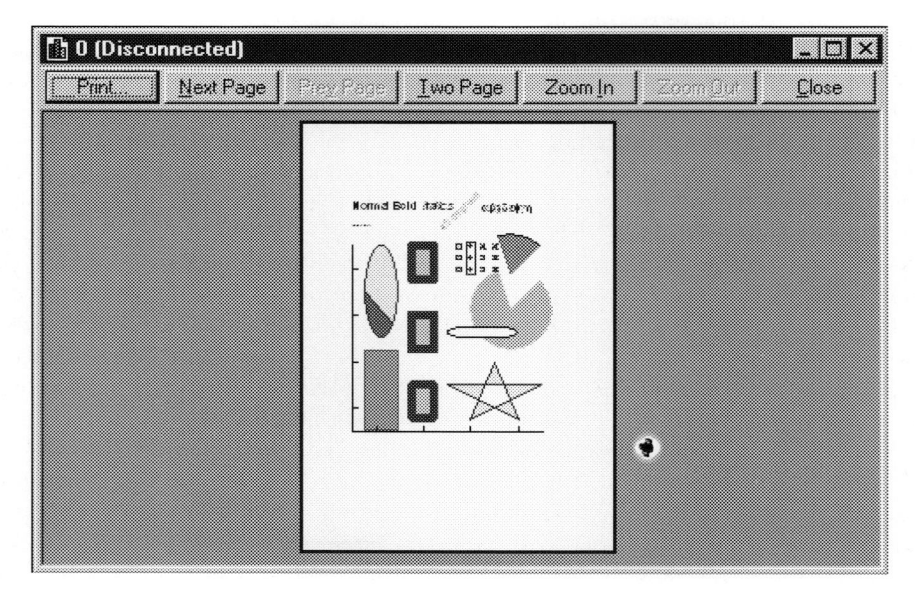

Fig. 3

Which format to choose?

The choice between bitmap or metafile format depends on the use you intend to make of the file. If you are planning to use the graph as a finished illustration, you can save it as a bitmap. The same thing applies if you are going to add a few things to the graph, like text labels, a title or a banner, but you are planning no change to pre--existing elements. It is practical to do this type of editing with a graphic editor. On the other hand, it is not possible to scale (i.e., change the size in pixels) of a bitmap without some loss in quality. If you reduce the size (in pixels) of a bitmap, some elements of the graph may disappear, become fuzzy, or become thinner or thicker than other elements. The exact result depends on the algorithm used to change the bitmap size, but there are no "perfect" algorithms. If you increase the size of a bitmap, some elements may become thicker or thinner than others and the whole graph may look "blocky" if observed from a close distance.

All graphic programs that can read images in Windows Bitmap format should yield the same results when displaying or printing the image at its original size (allowing for limitations of the output device, e.g., the number of available colours).

In conclusion, you should save a bitmap at the size determined by the use you intend to make of the graph. For instance, if you are planning to use the graph as an illustration 50 mm wide, and you know that the publisher will use a resolution of 1200 dpi (dots-per-inch), then you should save the bitmap with a width of 1200 * 50 / 25.4 = 2362 pixels.

A further limitation of the bitmap format is that Graph Server is limited to the range of colours that can be displayed on the screen of the computer. This applies also when saving a graph to a bitmap file. If your computer can display only a limited range of colours (e.g., 16 or 256) and you desire a broader range of colours in your graph, you must run Graph Server on a computer with better graphic capabilities, and save the bitmap on this computer.

The metafile format is not size- or resolution-dependent, and can be scaled freely without loss of precision. This format is often used to store libraries of ready-to-use images, generally referred to as clip-art. As a consequence, the metafile format is better known, among non-technical users, as clipart format. There are two metafile formats: an old one, called Microsoft Windows Metafile, inherited from Windows 3.1, which generally uses the **.wmf** file extension, and a more recent one, called Enhanced Windows Metafile, which uses the **.emf** file extension. Graph Server uses the latter format, because it stores more information about the original picture. The metafile format has the same colour limitations as a bitmap file (see above).

From the above discussion, it would seem that the metafile format is more flexible, and therefore preferable to a bitmap. However, although the metafile format is standard, graphic editors differ widely in the amount of information they extract from a metafile, and in the faithfulness of their rendering of a metafile. As shown below, the results can be extremely different. For instance, when confronted with a metafile generated from the graph at the left, a popular commercial program pro-

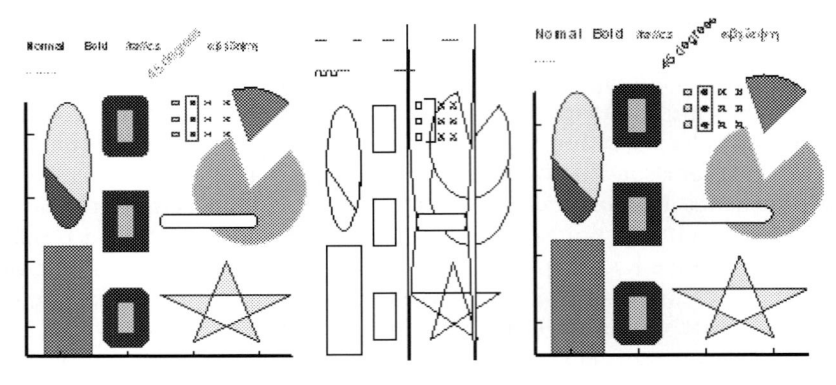

Fig. 4

duced the figure at the centre. Another popular program did considerably better, and produced the figure at the right. Note, however, that some details still differ from the original.

In conclusion, the advantages of a metafile are that (1) its file size is considerably smaller than a bitmap, (2) it can be scaled, and (3) once imported into a graphic program, its elements can be edited. For instance, it is possible to move or erase some of the elements in the foreground, thus making it possible to see underlying elements. Similarly, the contents of text labels can be edited, and the labels themselves can be moved. On the other hand, the disadvantage of metafiles is that it may be difficult to find a graphic editor that displays all graph elements in an acceptable way.

Saving as a bitmap

This option is available under the **File → Save as Bitmap** menu item. This command displays the following dialog box:

Fig. 5

Section 1. Using graph server 55

You should enter the desired vertical or horizontal size of the bitmap, expressed in pixels. When you enter a horizontal size, the corresponding vertical size is automatically computed, and vice versa. This prevents the bitmap from becoming "squashed" in the vertical or horizontal direction. By default, Graph Server will propose a horizontal size of 1000 pixels

The **Convert to monochrome** check-box is used to generate a monochrome bitmap. This is useful when a graph is to be used as a line illustration. All colours are eliminated, and converted to either black or white. In particular, light colours are converted to white, and dark colours to black. If the results do not match your expectations (i.e., some areas that should be white are rendered as black, and vice versa), try to lighten or darken the colour you are using for these elements (or make them black or white as appropriate).

Converting a large bitmap graph to monochrome reduces its file size, and allows other programs to load and process it more quickly.

Subsequently, Graph Server displays a standard **Save As** dialog box. You can use this dialog to "navigate" to the desired directory, and to specify a file name. If a file with the same name already exists, you will be prompted to specify whether you want to overwrite it.

After you press the **Save** button, Graph Server will prepare a copy of the bitmap in memory, and then save it to the specified file. If the bitmap is large, the process can be lengthy, and can fail to complete on computers with an insufficient amount of memory. The bitmap is saved to a file in the Windows Bitmap (**.bmp**) format. This format yields quite large files, but has the advantage of being universally understood by graphic editors.

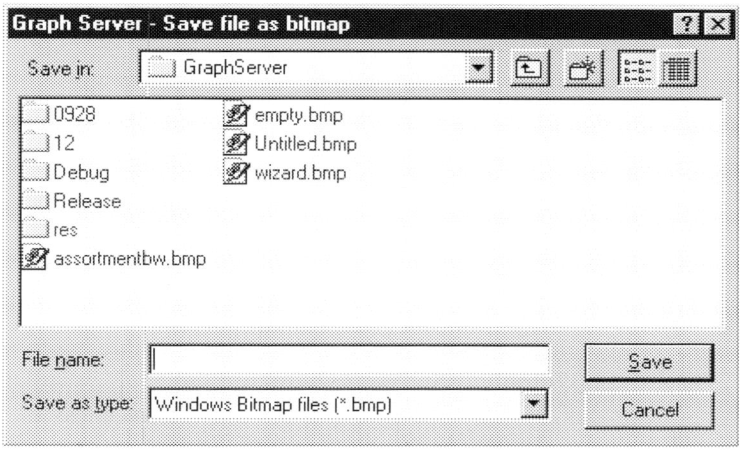

Fig. 6

Saving as a metafile (clipart)

This option, available under the **File → Save as Clipart** menu item, requires you to specify a file name and directory in a **Save As** dialog box. There are no other settings, because the metafile format is resolution-independent. See the above discussion on other merits and drawbacks of the metafile format.

SECTION 2. THE GS LANGUAGE

Scope of the GS language

The GS (acronym for Graph Server) language is the logical interface between Graph Server and programs written by users. Client programs run separately from Graph Server, and access its graphic capabilities by connecting to the server and issuing GS instructions. The user-interface of Graph Server allows users only to view, save and print the generated graphs. All other operations are performed through GS commands.

Design philosophy

The GS language wraps the Windows GUI in a device- and display-size-independent layer, thus avoiding the need to re-compute graphs when display sizes and output devices are changed. For instance, this allows Graph Server to view a graph on a colour screen, and subsequently print it on a high-resolution monochrome printer, without needing additional communication from the client.

   GS is designed to provide a simplified, but still powerful interface to the Windows graphical engine. Most of the low-level graphic commands native to the Windows GUI are accessible through this interface. In addition, GS expands the range of low-level graphical functions available in Windows (in particular by allowing the use of relative co-ordinates in addition to absolute ones), and adds a few high-level functions that are especially useful in graphing. GS is further enhanced by the capability of declaring and using symbolic values in addition to numerical and literal ones.

Scope of GS settings

Graph Server is a state-machine that stores all current settings internally. This means, for instance, that it is sufficient to set the colour, thickness and style of the drawing pen, and then use the pen for a series of drawing commands, without specifying the pen settings before each command.

Section 2. The gs language

The scope of settings is limited to the current graph. Settings are not stored between sessions, and do not affect the settings used by other clients that are simultaneously connected to Graph Server.

At the beginning of each session, all settings default to the values used by Windows device contexts. If you are in doubt about these values, you should state the settings explicitly at the beginning of each session.

GS units

Graph Server attempts to be device- and resolution-independent. All numerical values are stored in floating-point format. Therefore, it is possible to specify, for instance, a line thickness of 1.5 units. GS units are dimensionless, and their actual size in a displayed or printed graph depends on the size of the current viewport (see below). In a viewport with a size of 150 by 150 units, a line thickness of 1.5 corresponds to 1% of the viewport size. The actual size of a printout can be set in absolute units, or as a fraction of the page size. You can think of GS units as the measurement units you are familiar with. For instance, you may start by planning a graph size of 100 by 50 mm, and setting the viewport to 100 by 50 units. Major lines used in the graph can have a thickness of 0.5 mm, and minor ones 0.25 mm (consequently, you can use a line thickness of 0.5 and 0.25 units, respectively). Co-ordinates are Cartesian, and the y co-ordinate increases in the upward direction (not downward as in the Windows GUI).

The viewport

The viewport can be regarded as a rectangular window on the plotting surface used to draw a graph. While the plotting surface is infinite, the viewport spans the surface expressed by the co-ordinates of two points corresponding to opposite vertices of the viewport rectangle. Only the portion of plotting surface comprised within the viewport can be seen. The viewport can be changed at any time during a session without affecting a graph (except for changing which portion of it is visible). Consequently, the viewport can be used, for instance, to pan a window across a graph and display or print selected regions of it. This is the suggested way to split a graph across multiple printed pages.

The viewport is not the same as the document window in Graph Server. The document window can show the whole viewport, or part of it. While the viewport can be changed only through the *client* program, the document window can be sized and scrolled on the viewport only *within* Graph Server. The size and magnification of the document window do not affect printouts, while the viewport settings do.

Precision of displayed, printed and saved graphs

The equipment you are using may be unable to display graphic elements with the chosen attributes. When the chosen colour cannot be matched exactly by the display, the closest approximation compatible with the equipment (as decided by the Windows interface) will be used.

When the size of a graphic element cannot be displayed exactly, its closest approximation (as chosen by the Windows graphical engine) is used. When the closest approximation would be zero pixels, however, Graph Server forces the graphic element to take the size of one pixel. This prevents sub-pixel-sized elements, like very thin lines, from becoming invisible. A trade-off is that thin lines may be rendered differently when displayed on-screen and printed (because screen resolution is typically lower than a printer's). You must be aware that this is a limitation of screen displays and/or printers, not of Graph Server. You *can* make thin lines appear of the correct size on paper (within the limitations of the printer) by stating the appropriate line thickness in the GS commands.

Graphs saved to a file may have different (and typically lesser) restrictions on colours and sizes than those displayed on screen or printed on paper. Colours are always saved to a file with values of their red, green and blue components expressed as integers ranging from 0 to 255 (which accommodate about 16 million colours). The size in pixels of a saved image can be specified before saving, and can be different from the size of a displayed and printed image.

Computing and saving images at high resolution (e.g., 4000 by 4000 pixels and above) may be a slow process, and may require large amounts of RAM and disk space.

The placement of graphical elements by Graph Server is constrained by the error implicit in single-precision, floating-point variables.

Graph Server has no built-in limits to the size of a graph, other than the range of values that can be specified by single-precision floating-point values (approximately $3.4 \cdot 10^{-38}$ to $3.4 \cdot 10^{-38}$). On the other hand, Windows NT and Windows 95 have different limits in the range of co-ordinates (expressed in logical units, i.e. pixels) they can accept. Windows NT uses 32-bit integers, which accommodate a value range of 0 to approximately $4.2 \cdot 10^9$. Windows 95 at present uses 16-bit integers, with a range of 0 to 65,535. You should be aware that this last range is small enough to cause problems on high-resolution devices, if a large printout size is used. For instance, on a photo-setter with a resolution of 4800 DPI (dots-per-inch), the maximum printout size under Windows 95 will be less than 35 by 35 cm.

Text labels displayed at a small size in a graph window may be incorrectly positioned on the page, and/or incorrectly sized. This is a limitation of the Windows graphical engine, not of Graph Server, and is particularly apparent at small text sizes and/or on low-resolution displays. As a rule, zooming-in on the text will correct this problem. The same graph, displayed in the *Print → Preview* window, is generally displayed more correctly. Text in a printed graph is correctly sized and scaled.

There is a limit (about 32 KB) to the length of a statement which can be sent to Graph Server. You are unlikely to reach this limit in normal usage, but you must be aware of it if you try to create a very long string, or to declare a very large array of constants.

Structure of the GS language

GS is used by client programs to communicate with Graph Server. This language is not meant for generic programming. Therefore, all numerical processing must take place within the client program. However, Graph Server stores both data and function calls before carrying out the graphic commands, and GS provides syntactical elements to use this capability. This is the principal source of flexibility in Graph Server. For instance, screen displays and printers typically possess different resolutions and graphic capabilities. Switching from one type of device to the other requires Graph Server to re-compute a graph. Since Graph Server stores data and commands, switching output format (or changing display size and position) is accomplished without a need for the client to re-issue the whole series of commands.

The syntax of GS partly resembles that of the C programming language. However, GS is much simpler in that it possesses no mathematical operators, conditional constructs or user-definable functions. Users can extend the capabilities of GS by adding compiled libraries of functions. This requires familiarity with the C (or, preferably, C++) programming language and with the internal workings of Graph Server.

GS processes two categories of instructions: statements and comments.

GS statements

A GS statement consists of an alphanumeric string containing GS syntactical elements and ending with a semicolon. A GS statement must be completely specified in a single instruction. Instructions containing an incomplete statement, as well as instructions containing multiple statements, are regarded as syntax errors.

Leading and trailing spaces, tabs and carriage returns, as well as multiple spaces, tabs and carriage returns within statements, are ignored. Multiple spaces are retained, however, when they occur within string constants (see below). Spaces are not significant, and can be absent, between syntactical elements. The only exceptions to this rule are declarations of symbolic constants (see below). A space must be present between the type and the name of the constant, because this is the only way to tell the parser where one element ends and the next begins. The same rule applies to variable declarations in C and C++. Spaces are not allowed *within* a syntactical element.

GS comments

GS allows the use of comments within statements. Comments are ignored by Graph Server, but may be useful to document a sequence of instructions. A comment starts with the character pair // and continues to the end of the instruction. A comment can be the only content of an instruction:

// this whole instruction is a comment

Leading spaces preceding a comment are ignored. A comment can also be appended to a statement:

MoveTo(0, 0); // this is a comment appended to a statement

An instruction containing no character or only spaces is regarded as a comment.

Numerical constants

GS accepts numerical constants expressed in all common decimal formats. All following examples are acceptable:

-10
2.5E-1
.01
0.01

String constants

String constants are alphanumeric strings enclosed within quotation marks:

"Hello world!"

Characters that cannot be directly typed as part of a string constant (e.g., quotation marks) are entered by using escape sequences. Like in C and C++, escape sequences are character pairs starting with a backslash. The following escape sequences are defined in GS:

this escape sequence	evaluates to
\n	carriage return
\t	tab
\\	\
\"	"
\'	'

If a backslash is followed by an undefined escape character, the backslash is ignored and the character that follows it is entered in the string.

Special attention must be paid when writing a C or C++ program which sends a string containing escape sequences to Graph Server (see below).

Symbolic constants

In addition to numeric and string constants, GS allows the use of symbolic, i.e., named constants. These are constants associated with an alphanumeric name. Once declared and initialized, symbolic constants can be referred to by their name. This is especially useful with arrays (see below).

An important difference between GS and typical programming languages is that all symbolic names in GS are associated to constants, rather than variables. Therefore, the numerical or string values associated with a symbolic name cannot change once initialized. This is a consequence of the fact that GS is not meant to perform data processing.

The GS language provides three types of symbolic constants:

val – a floating-point value.
point – a point in two-dimensional space, expressed as an (x, y) co-ordinate pair of floating-point values.
string – an alphanumeric string of characters.

Symbolic constants must be declared and initialized within a single statement:

val i = 0;
val pi = 3.1415;
point centre = 10, 10;
// note the comma separating the (x, y) values
string MyName = "whatever you want";

In the argument lists of function calls (see below), the two numeric values contained in a **point** constant can be individually accessed as **centre.x** and **centre.y**. In this context, the point character is called *member operator*. Readers familiar with the C or C++ languages can regard the **point** type as a structure containing two floating-point members, **x** and **y**.

The general rules for symbol names are:

- Names must start with a letter character.
- Names can contain letter and number characters.
- Names cannot contain other characters (e.g., spaces, =, |, –).
- Names are case-sensitive: the names **num**, **NUM**, **Num** are regarded as different.

- Names of commands, symbolic constants, data types and functions must be unique. It is not possible to create a symbolic constant with the name used for a command, function or type, or with the name of another constant (of any type) that has already been declared.

Symbolic constants are accepted in place of numerical constants as arguments of function calls (see below). However, symbolic constants cannot be used to initialize other symbolic constants.

It is illegal to change the value associated with a symbolic constant, and to skip the initialization of a symbolic constant in its declaration. Each of the statements

val i; // no initialisation, value is undefined
i = 10; // attempting to change the value of a constant

constitutes an error in GS.

Arrays

Array declarations in GS are similar to declarations of the corresponding simple types, but the values used to initialize the array must be enclosed within braces. Unlike in C and C++, it is not necessary to specify the number of elements contained in the array being initialized. The number of elements is computed from the number of available initialization values.

val a = {1.0, 2.0, 3.0};
point pt = {0, 1, 2, 3};
// the values are: pt[0].x = 0, pt[0].y = 1, pt[1].x = 2, pt[1].y = 3

All elements of an array are initialized in the array declaration. Therefore, the declaration of a **point** array must contain an even number of initialization values. Arrays must contain at least one element. GS provides for arrays of **val** and **point** constants. String arrays are not implemented.

In function calls (see below), an element of an array can be accessed by enclosing its index within square brackets and appending it to the constant name (e.g., **b[1]**). Specifying the name of an array without index (e.g., **b**) indicates that the whole array is used (note that this aspect of the GS syntax differs from C and C++ array notation).

Element indices start at 0. Therefore, an array of 10 elements contains elements with indices ranging from 0 to 9 inclusive. Indices are integer values, and cannot be negative. It is an error to use an index exceeding the actual number of elements contained in an array. Array indices must be expressed in numerical form. Symbolic constants cannot be used as indices.

Functions

Functions are symbolic names that indicate an operation to be performed. Syntactically, a function consists of a name followed by a list of zero or more arguments enclosed within parentheses. Optional spaces can separate the function name from its argument list. Arguments are separated by commas and, optionally, spaces. Function names are case-sensitive. Examples:

Viewport (0, 0, 100, 50);
Title("My graph");

GS accepts values expressed as numeric constants, or string constants delimited by quotes (see examples above). Graph Server stores all numeric values (except for array indices) as single-precision floating-point numbers.

GS provides several tens of built-in functions. Function names follow the same syntactical rules of symbolic constant names. All built-in function names have capital initials, and it is recommended that the same rule be followed when expanding GS. Thus, using constant names with a lowercase initial avoids conflict with function names. In the following discussion, function prototypes identify the function arguments by type (in a manner analogous to C and C++ function prototypes). Type names are not used when calling functions. For instance, the function prototype of **Viewport()** can be written as:

Viewport(val x1, val y1, val x2, val y2);

This means that the function should be called with four arguments of type **val**.
In the above prototype, the arguments are given names that help to identify their usage in the description of the function. In actual use, you may call this function as:

Viewport(0, 0, 1, 1);
val a = 0;
val b = 1;
Viewport(a, a, b, b);
Viewport(a, 0, 1, b);

A function prototype with an empty argument list means that the function does not take any argument (e.g., **PushSettings( );**). The parentheses, however, must still be used in the function call.

GS implements two forms of function overloading. Function overloading is missing in C, where a function must be called with a set of arguments of predefined number and type (the only built-in way to change this behaviour involves completely defeating the type-checking mechanism). As implemented in C++, function overloading allows two or more functions to have the same name. Each of these functions accepts a different set of arguments, and the code of each function must be written separately. C++ compilers decide which overloaded function to call on the basis of the type and number of arguments specified in the function call.

Function overloading in GS, on the other hand, behaves differently. An overloaded GS function can be called with different numbers and types of arguments, but the same function is called in all cases. For instance, as discussed above, **Viewport** needs four **val**s as arguments. It is also possible to provide the arguments as two **point**s, a **point** and two **val**s, an array of four **val**s, an array of two **point**s, or any other combination of numeric and symbolic constants that can be reduced to four **val**s. It is illegal, however, to pass a **string** as an argument to **Viewport**. It is also an error to pass fewer or more than four **val**s.

The second form of overloading allows GS functions to accept a variable number of arguments. This is expressed in their prototypes with the use of ellipses that follow the minimum number of arguments. For instance, the prototype

LineTo(val x1, val y1, ..);

indicates that the function **LineTo** accepts two or more **val**s. Since this function accepts co-ordinate pairs, it is an error to use an odd number of arguments. Therefore, it is desirable to express this limitation in the function prototype. This can be done by using constants of type **point**, rather than **val**, as arguments:

LineTo(point p1, ..);

Function reference

This part of the documentation contains the description of all GS functions. On the CD-ROM, the function reference can be searched by categories of functions or alphabetically by function name. Examples of the use of most functions can be found in the GS tutorial (section 3 of this chapter).

General settings

Viewport

Viewport(point p1, point p2);

Sets the position and size of the viewport with respect to the current origin (see below). The viewport co-ordinates are expressed as the opposite vertices of a rectangle. See the discussion in "Settings, units and the viewport".

Section 2. The gs language

VScale

VScale(val f);

Sets the vertical scaling factor. All input *y* co-ordinates are multiplied by factor **f** before being displayed. This function is used to set-up an anisometric *xy* co-ordinate system. The default value of the vertical scaling factor is 1.

This function can be used to correct the distortion introduced by output devices that use rectangular, rather than square pixels.

Graphic settings

The functions listed below specify graphic settings. The settings remain valid until changed, and are used by all graphic functions.

PenThickness

PenThickness(val t);

Specifies the thickness of points and lines drawn with the current pen. A thickness lesser than or equal to 0 produces a line with a thickness of one pixel.

PenColor

PenColor(val red, val green, val blue);

Specifies the colour of the current pen. The colour is expressed by three values corresponding to the red, green and blue components of the colour. The value of each component can vary from 0 (darkest) to 1 (lightest). Values higher than 1 are set to 1. Values lesser than 0 are set to 0. The following examples show some common colours.

components	**colour name**
(0, 0, 0)	black
(1, 1, 1)	white
(1, 0, 0)	red
(0, 1, 0)	green
(0, 0, 1)	blue
(1, 1, 0)	yellow
(0.5, 0.5, 0.5)	medium grey
(0.14, 0.38, 0.11)	olive green

PenColorBytes

PenColorBytes(val red, val green, val blue);

Specifies the colour of the current pen. This function accepts red, green and blue colour components with values ranging from 0 to 255. These value ranges are the same used internally by Windows and by most graphic software. This function behaves like **PenColor**.

LineEndRound, LineEndSquare, LineEndFlat

LineEndRound();
LineEndSquare();
LineEndFlat();

These functions set the appearance of the extremities of line segments drawn by **LineTo** and **LineBy**. **LineEndRound** draws a semicircle centred on the end-points of the line. **LineEndSquare** draws a half-square centred on the end-points of the line. **LineEndFlat** causes the end-points to be drawn with a straight border perpendicular to the line direction. The default is **LineEndRound**.

LineJoinRound, LineJoinBevel, LineJoinMiter

LineJoinRound();
LineJoinBevel();
LineJoinMiter();

These functions set the appearance of vertices drawn with **LineTo**, **LineBy**, **Rectangle**, **Polygon**, **Chord**, **Pie** and **Marker**. **LineJoinRound** rounds the convex side of vertices. **LineJoinBevel** bevels the convex side of vertices (i.e., uses a straight-line segment to "cut out" the corner). **LineJoinMiter** miters the convex side of vertices (i.e., renders vertices as sharp corners). If the miter exceeds 10 pixels in length, the convex side is beveled. The default is **LineJoinRound**.

FillColor

FillColor(val red, val green, val blue);

Sets the fill colour (used, for instance, to fill a polygon). The colour components are expressed as values ranging from 0 to 1.

Section 2. The gs language   67

FillColorBytes

FillColorBytes (val red, val green, val blue);

Similar to **FillColor()**, but accepts values ranging from 0 to 255.

PenStyle

PenStyle(val on1, val off1, ..);

Sets the style of the current pen. The argument only specifies that initially the pen will draw a segment of the specified length. Then the pen will be lifted for the length specified by off1, and so on. When all arguments have been used, the pattern repeats.
   Call **PenStyle** with just one argument (of any value) to draw continuous lines. This is the default style.

ModeDefault, ModeNot, ModeOr, ModeAnd, ModeXor, ModeMerge

ModeDefault();

Sets a drawing mode in which new elements replace the pre-existing ones they overlap. This is the default mode. Other modes are set by the functions:

ModeNot();	colour components of the overwritten elements are inverted.
ModeOr();	colour components of the existing and new elements are logical OR-ed.
ModeAnd();	colour components of the existing and new elements are logical AND-ed.
ModeXor();	colour components of the existing and new elements are logical XOR-ed.
ModeMerge();	colour components of the existing and new elements are merged.

Simple graphic functions

The functions in this group draw lines and points with the current pen colour, thickness and style, or move the current drawing position.

MoveTo

MoveTo(point p);

Moves the current drawing position to the specified co-ordinates. It is equivalent to moving a pen while holding its tip lifted from the paper.

LineTo

LineTo(point p1, ..);

Draws a line segment with the current pen from the current drawing position to the specified co-ordinates. If several co-ordinate pairs are specified, each successive point is joined to the preceding one by a segment.

LineBy

LineBy(point p1, ..);

Similar to **LineTo**, but co-ordinates are relative to the last current position. For instance, the statement

LineBy(10, 0, 0, 10);

draws a straight line segment from the current position to a position 10 units to the right, then another segment from the new position to one 10 units upward.

Arc

Arc(point p1, point p2, point pStart, point pEnd);

Draws an elliptical arc. The ellipse is defined by the opposite vertices of its bounding rectangle (**p1**, **p2**). The starting and ending points are specified as the co-ordinates of points lying on ellipse radials (**pStart**, **pEnd**). The arc is drawn in the counter-clockwise direction.

Point

Point(point p1, ..);

Draws one or more points with the current pen at the specified co-ordinates.

Section 2. The gs language

PointBy

PointBy(point p1, ..);

Similar to **Point**, but co-ordinates are relative to the last-used ones.

Pixel

Pixel(point p1, ..);

Draws one or more pixels with the current pen colour at the specified co-ordinates. **Pixel** always paints a single pixel at the specified co-ordinates, and does not use the current pen thickness. This is the only GS function that uses device-dependent units.

Filled shapes

The functions in this category use the current pen colour, thickness and style to draw a shape's border, and the current fill colour and fill mode to fill its inside area.

Rectangle

Rectangle(point p1, point p2);

Draws a rectangle specified by two opposite vertices.

Polygon

Polygon(point p1, point p2, point p3, ..);

Draws a polygon with the vertices specified as arguments. The first and last vertices specified in the argument list are automatically joined with a line segment.

Ellipse

Ellipse(point p1, point p2);

Draws an ellipse specified by two opposite vertices of its bounding-rectangle.

Pie

Pie(point p1, point p2, point pStart, point pEnd, val direction);

Draws an ellipse sector (or pie slice). The ellipse is specified by two opposite vertices of its bounding-rectangle (**p1, p2**). The start and end points of the pie along the perimeter of the ellipse are specified as the co-ordinates of two points located on ellipse radials (**pStart, pEnd**). If **direction** has a value of zero, the pie is drawn in the counter-clockwise direction, clockwise if non-zero.

Chord

Chord(point p1, point p2, point pStart, point pEnd, val direction);

Similar to **Pie**, but draws a shape enclosed between an elliptic arc and its chord.

Advanced graphic functions

These functions are specifically designed to help in the creation of graphs.

Marker

Marker(point p1, point p2, ..);

Draws a polygon with the vertices specified by the argument list. The co-ordinates in the argument list are interpreted as relative to the pen position at the time the function was called (*not* as relative to those of the preceding vertex).
   This function is useful for displaying a symbol or figure at the current drawing position. By using an array constant as the argument to **Marker** and alternating calls to **Marker** with calls to **MoveTo**, it is easy to repeatedly draw a figure at the co-ordinates of each data point in a set. For instance the statements:

point square = {-1, -1, -1, 1, 1, 1, 1, -1};
MoveTo(0, 0);
Marker(square);
MoveTo(5, 0);
Marker(square);
MoveTo(10, 0);
Marker(square);

draw three squares with a side of 2 units at the co-ordinates (0, 0), (5, 0) and (10, 0). The first and last points of a marker's sequence of vertices are automatically joined, and the polygon is filled with the current fill colour. To draw an open, non-filled line, use the **MarkerLine** function.

MarkerLine

MarkerLine(point p1, point p2, ..);

Similar to **Marker**, but draws an open, non-filled line. The first and last vertices are not automatically joined. If you want to draw a figure consisting of two or more disconnected series of segments, you can call **MarkerLine** multiple times. You can use the same method to draw a symbol consisting of intersecting lines. For instance, a '+' symbol can be drawn as:

MarkerLine(0, 0, 1, 0);
MarkerLine(.5, -.5, .5, .5);

Alternatively, you may retrace your path back to the intersection of two segments and continue from there in a new direction:

MarkerLine(0, 0, .5, 0, .5, .5, -.5, .5, 0, 1, 0);

Text functions

The following functions control the appearance of text. A detailed discussion and examples are provided in the tutorial (section 3 of this chapter).

FontName

FontName(string name);

The font specified by the argument becomes the current font. A run-time error occurs if a font with this name is not installed, and the current font is not changed.

FontHeight

FontHeight(val s);

Sets the current font size in GS units. The font size corresponds to the distance between the top of an uppercase letter and the bottom of a descender letter (i.e., a letter, like 'g' and 'q', that descends below the text baseline). In practice, font size is often defined as the difference in height between the top of an uppercase 'M' and the bottom of a lowercase 'g'. The actual size of a displayed font is an approximation, dictated by limitations of the Windows GUI and/or of the current font.

Font sizes are commonly expressed in points. A point is approximately 1/72 inch, or .35 mm. Since GS font sizes, like all other sizes, are based on dimensionless GS units, the actual size of a font depends on the size of a printed or displayed graph (see the section entitled "Settings, units and the viewport").

FontNormal, FontBold, FontItalic, FontUnderline, FontStrikeOut

FontNormal();
FontBold();
FontItalic();
FontUnderline();
FontStrikeOut();

These functions control the style of the current font. **FontNormal** resets all attributes to normal (i.e., non-bold, non-italic, non-underlined, non-stroked-out). The remaining functions set the attribute indicated by their name. You can combine two or more attributes by calling the appropriate functions consecutively.

FontAlignTop, FontAlignBase, FontAlignBottom

FontAlignTop();
FontAlignBase();
FontAlignBottom();

These functions cause text to be vertically aligned with respect to the current drawing position. **FontAlignTop()** aligns text in correspondence of the top of a bounding-box containing the text string. **FontAlignBase()** aligns text in correspondence of its baseline (i.e., the bottom of a line of text that does not contain descenders like **p** and **g**). **FontAlignBottom()** aligns text in correspondence of the bottom of its bounding-box (i.e., the bottom of the lowest descender).

FontAlignLeft, FontAlignCenter, FontAlignRight

FontAlignLeft();
FontAlignCenter();
FontAlignRight();

These functions align text horizontally with respect to the current drawing position.

FontAlignUpdate, FontAlignNoUpdate

FontAlignUpdate();
FontAlignNoUpdate();

FontAlignUpdate() causes the current drawing position to be moved to the end of displayed text. This allows new text to be output as a continuation of the same line. **FontAlignNoUpdate()** restores the default condition, in which text output does not change the current position.

FontInclination

FontInclination(val a);
Rotates text counter-clockwise by angle **a**, expressed in degrees.

Text

Text(string t);

Displays a string at the current drawing position. The text is aligned with the lower-left corner of its bounding-box at the current drawing position.

SECTION 3. DEBUGGING FACILITIES OF GRAPH SERVER

You can examine the statements and symbolic constants stored by Graph Server at any time during a session. To examine the symbolic constants, select the menu item *View → Data*. This displays a window containing a list of five items (Fig. 7).

Each item corresponds to a data type. If a small square button containing a **+** symbol appears at the left of an icon, one or more constants of the corresponding type have been declared during the current session. To expand a list, click with the left mouse button on the corresponding **+** symbol. When a list is expanded, the symbol in the button turns to a -. Click on this symbol to collapse the list.

The items contained in a data-type list contain the symbolic name of the data item. Each data item contains a list of the numerical or string values associated with the data item (Fig. 8).

To examine the stored function calls, select the menu item *View → Function Calls*. This displays a window containing a list of the stored function calls (Fig. 9).

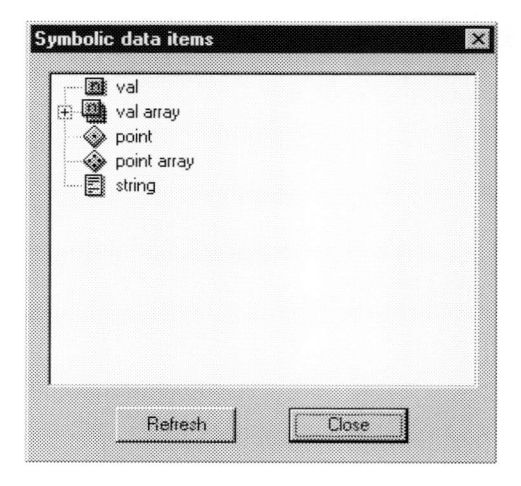

Fig. 7

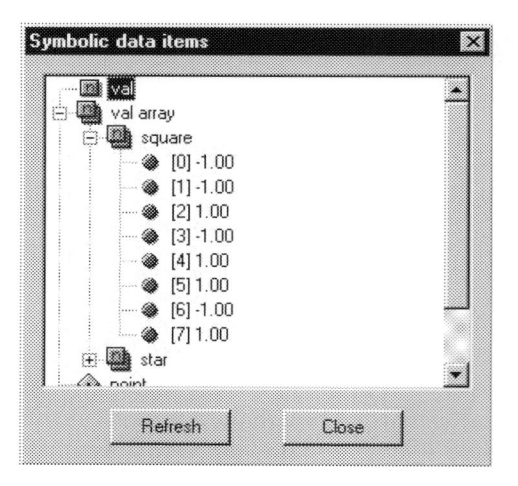

Fig. 8

Each function call contains a list of the numerical or string data received as arguments of the function. Note that symbolic arguments are translated to their numerical or string values in this list (Fig. 10).

If a run-time error is detected, the icon of the corresponding statement changes to a crossed-out tag, and the list associated with the offending statement contains an error message, in addition to the function arguments.

Note that the original GS statements are not stored by Graph Server, so comments and blank lines do not appear in the list of function calls.

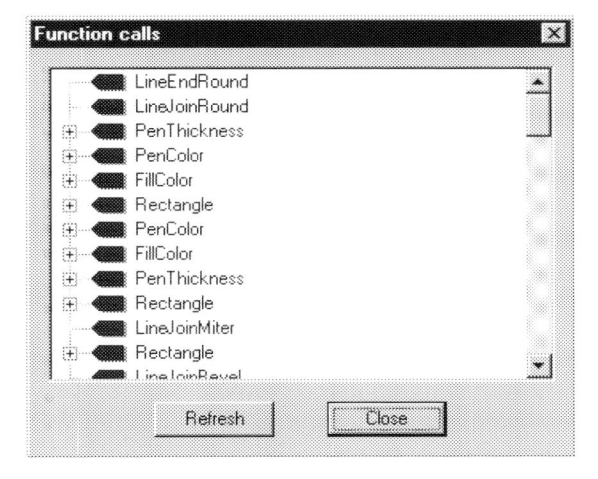

Fig. 9

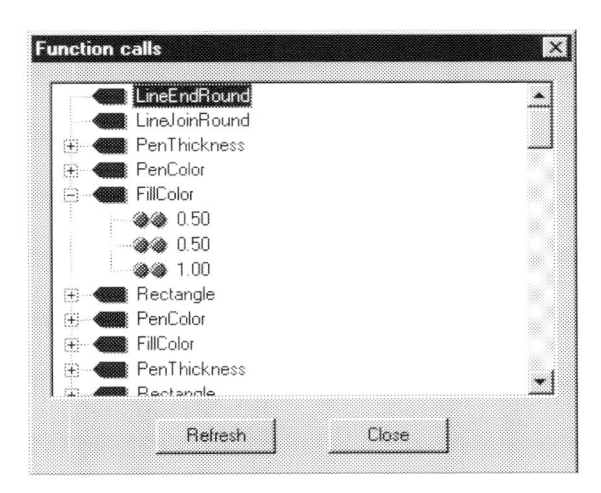

Fig. 10

SECTION 4. A GS TUTORIAL

This section provides a step-by-step introduction to using the GS language. This tutorial assumes that you already know how to interface your program to GraphClient.dll or, alternatively, that your program saves a sequence of GS commands to a text file, for later use by SendFile.exe. Both techniques are described in section 1 of this chapter.

When you intend to manually draw a graph on paper, you start by deciding the type of diagram you want (e.g., scatter-plot, histogram, or pie). Subsequently, you decide how big your diagram will be on the paper, and which range of values to use for the axes (a program can decide the last item by examining the data set it will plot). You then draw the axes, tick-marks, and data points (or other graphical objects representing the data), and finally add labels, captions and other text. The process with Graph Server is largely similar, but there are two fundamental differences between the two methods (see below). Both manual drawing and the use of Graph Server require careful planning.

The first difference with manual drawing is that Graph Server gives you freedom on the size and placement of the graph on the drawing surface. For practical purposes, you can consider a graph generated by Graph Server as an object that has been drawn on a paper sheet of infinite size. You can magnify any portion of a graph without loosing detail (within the precision limits of the constants used by Graph Server). The display or print area is a rectangular "window" on the graph that you can freely change during and after the graph is being generated.

The second difference is that Graph Server may require you to draw graph elements in an order that differs from the procedure used for drawing manually. Normally, when Graph Server draws a graph element, it erases all previous elements occupying the same position (i.e., earlier elements are "overwritten"). Therefore, if your graph contains overlapping objects, you must arrange the corresponding GS statements in the appropriate sequence. Advanced users can force Graph Server to behave differently, and use a number of logical operators to combine the new element with pre-existing ones (see the **Mode...()** functions in the GS language reference, in the preceding section).

Text labels in the illustrations of this section may appear "soft" or "fuzzy". This is a consequence of the fact that the illustrations are bitmaps captured from the screen display. Windows NT (and Windows 95 with add-on software) can use anti-aliasing in order to obtain a better visual presentation of on-screen fonts. This results in soft edges of the font characters. Text on printouts, on the other hand, is not anti-aliased but uses the resolution of the printer, which is typically high enough to make aliasing effects (i.e., stair-step effects on oblique edges) unnoticeable.

Example 1 – Drawing a histogram

In the following discussion, we solve the task of drawing a bar diagram, discussing the successive steps of the procedure. We assume that we want to draw a histogram containing four data classes (i.e., four bars). The numerical (i.e., height) values for the bars are 16, 48, 36, 12. Each bar will be filled with a different colour. The outline of each bar will be drawn with a thin black line. The y-axis will have

Section 4. A gs tutorial

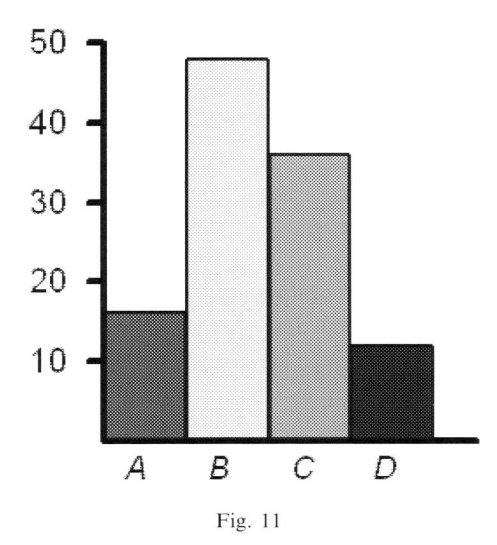

50
40
30
20
10

A B C D

Fig. 11

external tick marks at the values 10, 20, 30, 40, and 50. Each tick-mark is to be flanked by a label with the corresponding numerical value. The *x*-axis will carry the labels *A*, *B*, *C*, *D*, located directly beneath the axis, in correspondence of the bars.

The desired graph is shown as Fig. 11.

Drawing the axes and tick marks

Except where noted otherwise, the following discussion shows the GS statements that should be sent to Graph Server, but does not deal with the problem of how to generate these statements. Normally, GS statements are not manually written and issued one-by-one to Graph Server (although this approach *does* work). It is up to you to write a program, in the language of your choice, that generates the required statements. If your program is carefully written, it will require little or no modification to deal with different data sets.

At this stage, we set the origin of the plot at the (10, 10) co-ordinates, in order to leave space for the labels at the left of and below the axes. We do not use a scaling factor (i.e., we let Graph Server use the default scaling factor of 1). The *y*-axis will therefore range from 10 to 60 GS units. A suitable thickness for the axes is 1 unit. We start by setting the pen colour to black and the pen thickness to 1.

PenColor(0, 0, 0);
PenThickness(1);

We move the pen to the origin, and subsequently draw a line from the origin of the axes, i.e., (10, 10) to co-ordinates (10, 60):

MoveTo(10, 10);
LineTo(10, 60);

The tick-marks are supposed to be external, i.e., to project from the left side of the axis. The pen is now located at the last-used position, i.e., (10, 60). From there, we will draw the topmost tick in the left direction. A length of 2 units is reasonable. Now we can draw the topmost tick-mark:

LineTo(8, 60);

Subsequently, we draw the other tick-marks:

MoveTo(10, 50);
LineTo(8, 50);
MoveTo(10, 40);
LineTo(8, 40);
MoveTo(10, 30);
LineTo(8, 30);
MoveTo(10, 20);
LineTo(8, 20);

For the sake of efficiency, your program can generate repetitive statements within a loop. For instance, a C program can generate the required ticks as:

```
int y;
char c[256];
int left = 8, right = 10;
for(y = 20; y <= 60; y += 10)
{
    sprintf(c, "MoveTo(%d, %d);\n", left, y);
    send(c);
    sprintf(c, "LineTo(%d, %d);\n", right, y);
    send(c);
}
```

where the function **send()** issues commands to Graph Server (via GraphClient.dll), or saves them to a file for subsequent transmission.

Drawing the coloured bars

We decide to use a width of 10 units for each histogram bar, and to use a length of 45 units for the x-axis. In this way, the x-axis will project slightly from the rightmost end of the set of bars. We can now draw the x-axis:

MoveTo(10, 10);
LineTo(55, 10);

It is better to use a different line thickness for the outline of the bars, in order to set them off from the axes. We can use a thickness of 0.5.

PenThickness(0. 5);

The first bar will be filled with red.

FillColor(1, 0, 0); // Red

Since the bar is a rectangular shape, we can draw it with the **Rectangle()** command.

Rectangle(10, 10, 20, 26);

We subsequently draw the remaining bars, specifying a new colour for each.

FillColor(1, 1, 0); // Yellow
Rectangle(20, 10, 30, 58);
FillColor(0, 1, 0); // Green
Rectangle(30, 10, 40, 46);
FillColor(0, 0, 1); // Blue
Rectangle(40, 10, 50, 22);

Adding text labels

We are now ready to add the text labels. GraphServer uses the same text alignment of the Windows graphical engine. This means that text is written *to the left of* and *below* the current drawing position.

We further decide to use the Arial typefont with a bold style. Lastly, we need to set a font height (i.e., size). In the Microsoft Windows environment, font sizes are specified in points (1 point = 0.353 mm). Font size is defined as the height between the tallest ascender (e.g., the topmost portion of a capital **A**) and the lowest descender (the lowermost reach of a lowercase **g**). Graph Server, instead, uses GS units to measure font height (1 point = 0.353 GS units). This means that the font size of text labels as displayed on-screen or printed on paper depends on the scaling factor used for viewing a graph (i.e., text in a GS graph is scaled up or down, together with all other graph elements). If you must use a precise font size, you should plan in advance carefully, so that the scaling factor does not need to change to accommodate for bigger- or smaller-than-expected graphs. Alternatively, you can generate a graph without labels, and subsequently import it into a graphic editor for manual addition of the text labels.

A font height of 5 units (i.e., 10% of the height of the diagram) is appropriate for the labels.

FontName("Arial");
FontHeight(5);

The text labels alongside the tick marks can be added at this point. Since characters are 5 units high, it is necessary to place the current drawing position 2.5 units higher than the tick marks (remember that text is drawn *to the left of* and *below* the current drawing position). In a font with a height of 5 units, characters are, on average, roughly 3 units wide: as a rule, characters are slightly higher than wide, although the width of characters varies within the same fonts (e.g., compare the relative widths of the characters **M** and **I**). Since the *y*-labels consist of two digits, we can assume a total width of about 6 units. Therefore, we can add to this width one more unit (to provide a space between label and tick mark, and place the current drawing position 7 units from the left extremity of the tick mark (i.e., 9 units from the *y*-axis).

MoveTo(1, 22.5);
Text("10");
MoveTo(1, 32.5);
Text("20");
MoveTo(1, 42.5);
Text("30");
MoveTo(1, 52.5);
Text("40");
MoveTo(1, 62.5);
Text("50");

The text labels under the data bars are to be formatted in *italics*. Since these labels consist of a single character, we place the current drawing position 1 unit below the *x*-axis, and 2.5 units to the left of the centre of the corresponding histogram bar (we can assume a character width of 4–5 units, instead of 3, because italic characters are slanted).

FontStyleItalics();
MoveTo(12.5, 9);
Text("A");
MoveTo(22.5, 9);
Text("B");
MoveTo(32.5, 9);
Text("C");
MoveTo(42.5, 9);
Text("D");

Section 4. A gs tutorial

Choosing a viewport and displaying the graph

Finally, before we can display the graph we must specify a viewport, i.e., the size and placement of an observation window onto the infinitely large graphing-surface. We know that the graph, including its axes, extends between the co-ordinates (10, 10) and (55, 60). Allowing for the labels and a white margin around the whole graph, we set:

Viewport(0, 0, 70, 70);

At this point, we are ready to inspect the graph. We issue the command

Disconnect();

to detach our client program from Graph Server and allow the latter to generate the graph. The result is visible below.

Correcting the errors

This is the graph generated by the GS commands discussed in the previous sections:

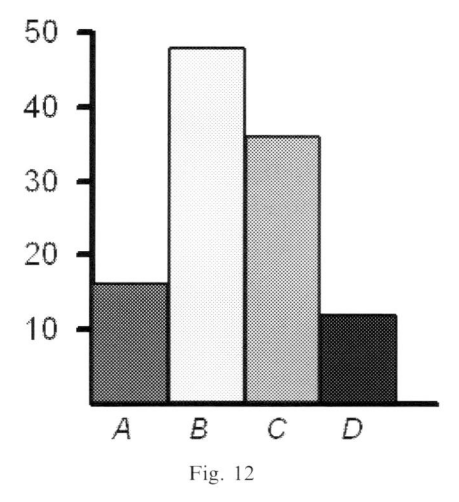

Fig. 12

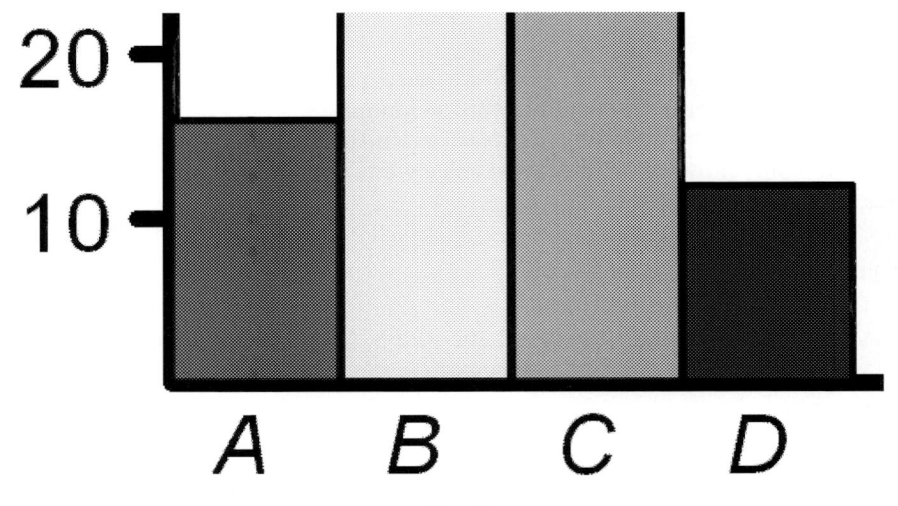

Fig. 13

A careful inspection of the graph shows that everything is as expected, with one exception: the colour fillings of the histogram bars partly overlap the x- and y-axes of the graph. This is especially visible when the graph is observed with a high zoom-ratio. In particular, the lowermost portion of the diagram is as shown in Fig. 13.

Compare the thickness of the y-axis above and below the top of the red bar. Unless a high zoom-ratio is used, this type of problem is likely to become apparent only when a graph is printed on a high-resolution device.

An examination of the GS statements above shows that the bars are drawn after the axes. Since the outline of the bars is thinner than the axes (0.5 versus 1 units), the colour filling of the bars partly overwrites the x- and y-axes. This can be corrected by re-arranging the sequence of GS statements in order to draw the axes after the bars. The final sequence, which corrects the above problem, will look like this:

Viewport(0, 0, 70, 70);
PenColor(0, 0, 0);
// draw the colour-filled bars
PenThickness(0. 5);
FillColor(1, 0, 0);
Rectangle(10, 10, 20, 26);
FillColor(1, 1, 0);
Rectangle(20, 10, 30, 58);
FillColor(0, 1, 0);
Rectangle(30, 10, 40, 46);

Section 4. A gs tutorial

```
FillColor(0, 0, 1);
Rectangle(40, 10, 50, 22);
// draw the y-axis
PenThickness(1);
MoveTo(10, 10);
LineTo(10, 60);
// draw the tick marks
LineTo(8, 60);
MoveTo(10, 50);
LineTo(8, 50);
MoveTo(10, 40);
LineTo(8, 40);
MoveTo(10, 30);
LineTo(8, 30);
MoveTo(10, 20);
LineTo(8, 20);
// draw the x-axis
MoveTo(10, 10);
LineTo(55, 10);
// draw the y-labels
FontName("Arial");
FontHeight(5);
FontStyleNormal();
MoveTo(1, 22.5);
Text("10");
MoveTo(1, 32.5);
Text("20");
MoveTo(1, 42.5);
Text("30");
MoveTo(1, 52.5);
Text("40");
MoveTo(1, 62.5);
Text("50");
// draw the x-labels
FontStyleItalics();
MoveTo(12.5, 9);
Text("A");
MoveTo(22.5, 9);
Text("B");
MoveTo(32.5, 9);
Text("C");
MoveTo(42.5, 9);
Text("D");
```

As a rule, the process of designing a new graph will be, in part, a trial-and-error procedure. The amount of work involved in programming a client application for Graph Server can be reduced by allowing your program to be flexible (e.g., by automatically choosing the plotting interval on the basis of the available data set). In this way, client programs can reuse common "building-blocks" that generate frequently-used types of diagrams.

Example 2 – A scatter diagram

We assume that we must generate a scatter diagram containing the following data points, identified by their (x, y) co-ordinates:

(12, 34) (41, 37) (25, 17) (8, 18)

Unlike the preceding example, we place the origin of the graph at the (0, 0) co-ordinates. In this way, we can directly use the co-ordinates of the points to be plotted. We can include the origin (0, 0) in the diagram. The maximum range of the co-ordinates used by the data points is (41, 37). We can therefore extend the axes to the co-ordinates (45, 40). To allow for labels at the left of and below the axes, and for a margin above and to the right of the diagram, we set the viewport at:

Viewport(0, 0, 50, 55);

Since the data points are far from either axis, we can draw the axes at this point.

PenColor(0, 0, 0);
PenThickness(1);
// draw the axes
MoveTo(0, 40);
LineTo(0, 0);
LineTo(45, 0);

We decide to mark each data point by drawing a marker centred at the co-ordinates of the point. We will use the same marker (a blue square with a yellow fill-colour) for all points. Therefore, we can define a marker shape, and subsequently re-use it. To do this, we store the co-ordinates of the corners of a square, centred about the origin, in an array, to which we assign the symbolic name *square*.

point square={-1,-1, 1,-1, 1,1, -1,1};

Section 4. A gs tutorial

We subsequently set the style of line joints to *miter* (e.g., sharp corners):

LineJoinMiter();

Now we are ready to display the markers.

PenThickness(0.5);
PenColor(0, 0, 1);
FillColor(1, 1, 0);
MoveTo(12, 34);
Marker(square);
MoveTo(41, 37);
Marker(square);
MoveTo(25, 17);
Marker(square);
MoveTo(8, 18);
Marker(square);

If we open the *View → Data* window in Graph Server, we can verify that the symbolic constant *square* has been stored (Fig. 14).
 Similarly, the *View → Function calls* window shows all the above function calls. Note that the arguments of the calls to **Marker()** have been changed to the numeric values associated with the symbolic constant (Fig. 15).

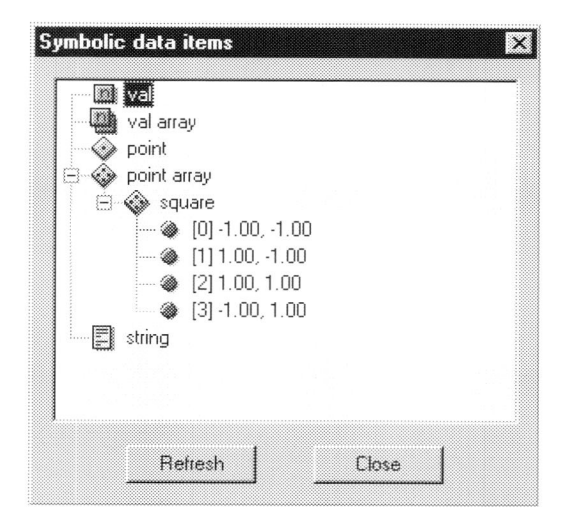

Fig. 14

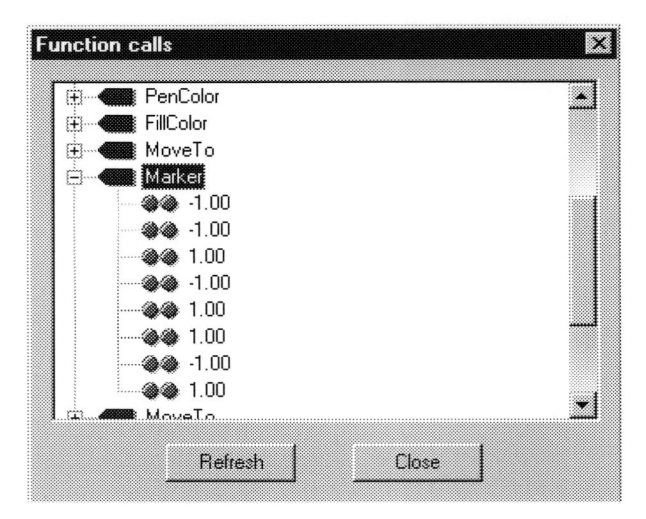

Fig. 15

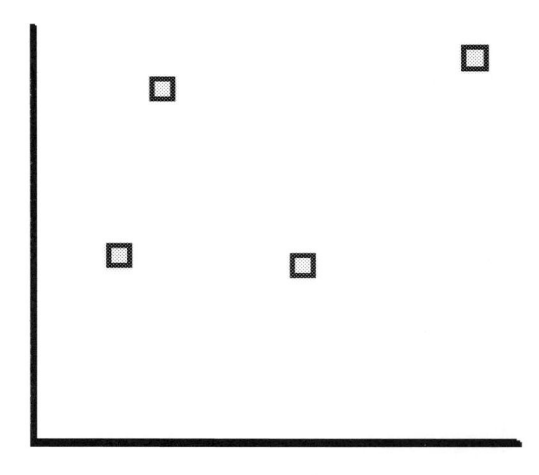

Fig. 16

The graph is displayed as Fig. 16.

An unexpected feature of this graph is that half of the thickness of the axes is missing. This is due to the fact that the axes start at the (0, 0) co-ordinates. Therefore, half of the axes thickness lies in regions where either x or y (or both) are negative. These regions are not displayed because the lower-left corner of the viewport is

Section 4. A gs tutorial   87

located at the co-ordinates (0, 0). Therefore, the margins of the viewport bisect the axes along their length. We can correct this problem by choosing a viewport that includes negative values:

Viewport(-5, -5, 50, 55);

At this point, we could add scale-marks and labels as discussed in the preceding example. However, GS provides better text facilities than those used so far. This is the subject of the next example.

Example 3 – Advanced text facilities

As discussed above, GS conforms to the default Windows behaviour when displaying text. However, it is possible to change this behaviour. In graphs, it is often useful to display text that is aligned to an axis different from the horizontal. For instance, the label of a *y*-axis is conveniently rotated 90 from the horizontal. Text rotation is achieved by the **FontInclination()** function. The figure below displays the result of the following GS code:

```
FontName("Arial");
FontHeight(10);
MoveTo(50, 50);
PenThickness(3);
Point(50, 50);
Text("text");
FontInclination(90);
Text("text");
FontInclination(180);
Text("text");
FontInclination(270);
Text("text");
```

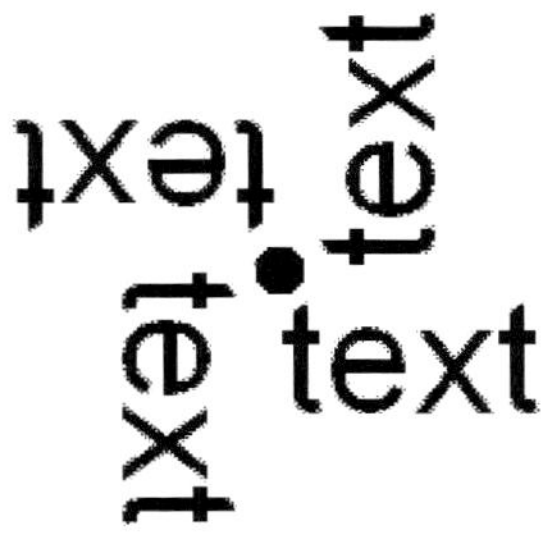

Fig. 17

In the above example, text is plotted at the current drawing position, which is marked by a black dot (generated by the **Point()** function). Note that rotation is measured in degrees, and in a counter-clockwise direction. Note also that the current drawing position is not changed by successive calls to the **Text()** function.

In the first example of this tutorial, text labels were positioned with respect to other elements of a graph by estimating the height and width of the text. GS provides a better way (albeit not a perfect one). In the following examples, the current drawing position is marked by the intersection of two red lines.

The function **FontAlignTop()** aligns text with the top of a bounding-box surrounding the text at the current drawing position. This is the default alignment (Fig. 18).

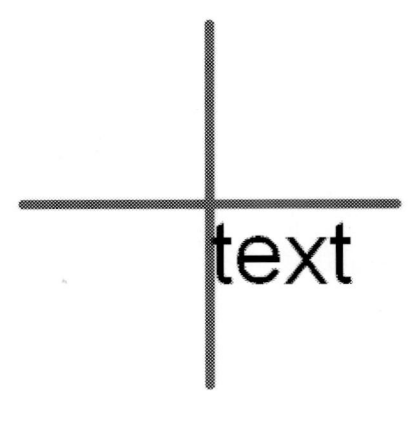

Fig. 18

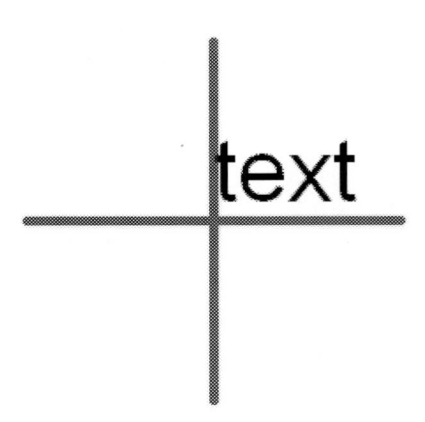

Fig. 19

The function **FontAlignBottom()** aligns the bottom of the bounding-box to the current drawing position. Note that the bounding-box is high enough to contain descenders, i.e., characters, like **p** and **g**, that descend below the baseline of the text (Fig. 19).

FontAlignBase(), on the other hand, aligns the baseline of the text to the current position (Fig. 20).

FontAlignLeft() (which is the default), **FontAlignCenter()** and **FontAlignRight()** control the horizontal alignment of text (Fig. 21).

A legitimate question is how alignment is set for text that has been rotated. The answer is that the result is equivalent to rotating the displayed text about the

Fig. 20

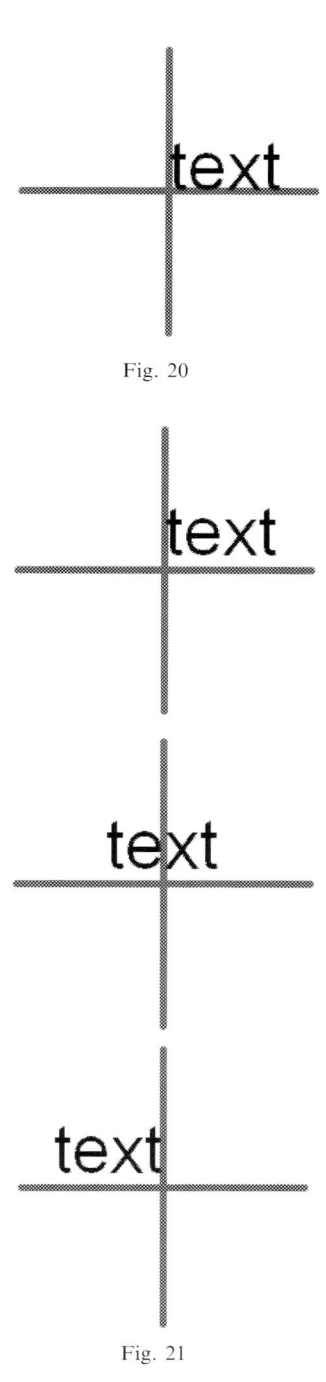

Fig. 21

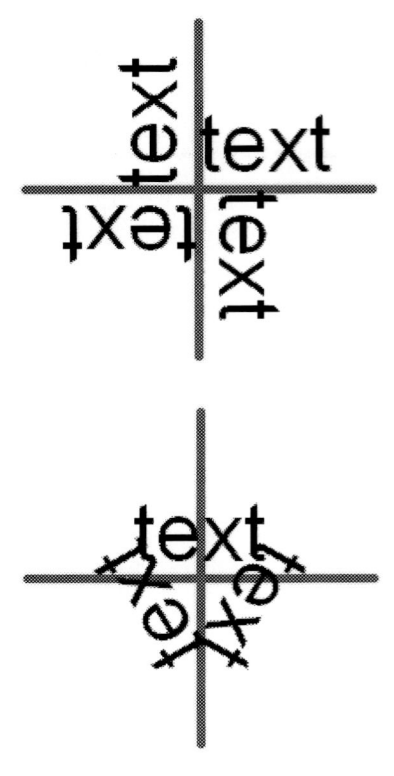

Fig. 22

current position. The figure below shows the effects of rotating left- and bottom-aligned text, and centre- and bottom-aligned text, respectively, by varying amounts (Fig. 22).

Unfortunately, the Windows graphical engine does not provide a simple way to centre text vertically (principally because there are many different ways to define this operation, i.e., centring about uppercase versus lowercase text, including or excluding ascenders and/or descenders). Doing so in GS does require the user to estimate the height of the text, and to choose one of the above ways to centre text vertically. The simplest way is probably to halve the nominal height of the text font.

We encounter another problem when we want to place two or more text labels to form a single line of text. This may be desirable when we must write a line of text that contains one or more symbols (or other characters that necessitate a change of font). For instance, we want to write $\pi = $ **3.14**. The symbol π is contained in a different font (e.g., Microsoft Symbol) than normal text. The function

Section 4. A gs tutorial

$$\pi = 3.14$$

Fig. 23

FontAlignUpdate() provides a way to solve this problem. Once this function is called, each call to **Text()** moves the current position to the end of the displayed text. The following example shows the result of the code (Fig. 23):

```
FontName("Symbol");
FontHeight(10);
MoveTo(30, 50);
PenColor(0,0,0);
FontAlignUpdate();
Text("p");
FontName("Arial");
Text(" = 3.14");
```

Note that the symbol π must be written, in GS code, as its equivalent character (i.e., the character that occupies the same position in the font table, and has the same ASCII value) in a normal font. Without the call to **FontAlignUpdate()**, the result would be as shown in Fig. 24.

Because the current drawing position would not move after drawing the π symbol. This would have to be corrected by estimating the width of the symbol, and moving the current drawing position by an appropriate amount. The effect of **FontAlignUpdate()** is reversed by **FontAlignNoUpdate()**.

Text can be formatted by using the functions **FontItalics()**, **FontBold()**, **FontUnderline()** and **FontStrikeOut()**. All these functions can be combined together by calling them consecutively. The function **FontNormal()** cancels all their effects.

$$\pi = 3.14$$

Fig. 24

This sequence of calls:

```
FontName("Arial");
FontHeight(6);
PenThickness(1);
MoveTo(5, 50);
PenColor(0,0,0);
FontAlignBottom();
FontAlignLeft();
FontAlignUpdate();
Text("normal ");
FontUnderline();
Text("underlined ");
FontNormal();
FontStrikeOut();
Text("struck-out ");
FontUnderline();
Text("both");
MoveTo(5, 40);
FontNormal();
FontBold();
Text("bold ");
FontNormal();
FontItalics();
Text("italics ");
FontBold();
Text("bold italics ");
FontUnderline();
FontStrikeOut();
Text("all");
```

generates the following text:

Fig. 25

Section 4. A gs tutorial

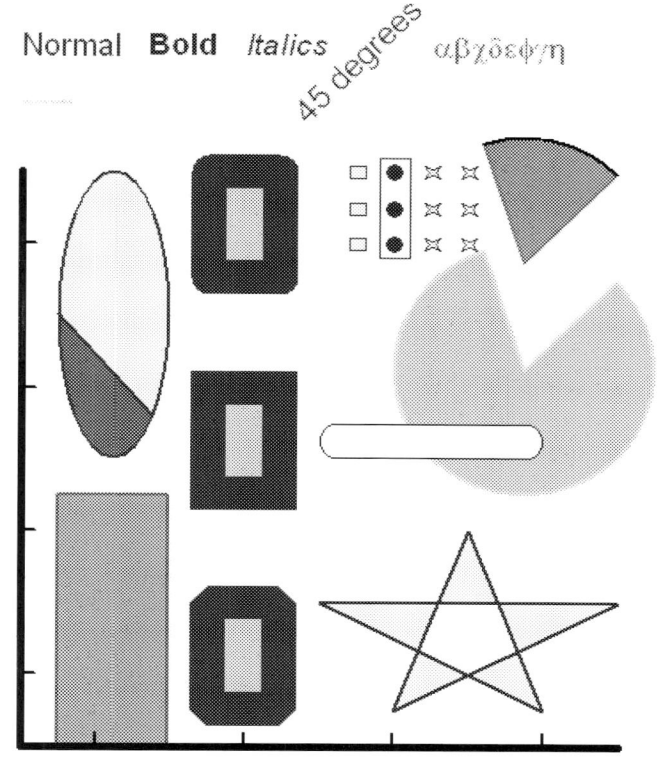

Fig. 26

Example 4 – An assortment of graphic and text effects

The above graph (Fig. 26) contains an assortment of graphic and text effects. This is the GS code that generates this graph:

Viewport(0, 0,100,120);
LineEndRound();
LineJoinRound();
PenThickness(.5);
PenColor(1,0,0);
FillColor(.5,.5,1);
// examples of rectangles
Rectangle(15,10,30,45);

```
PenColor(0,.2,.2);
FillColor(.8,.8,0);
PenThickness(4);
Rectangle(35,75,45,90);
LineJoinMiter();
Rectangle(35,45,45,60);
LineJoinBevel();
Rectangle(35,15,45,30);
// axes with ticks
PenThickness(1);
PenColor(0,0,0);
MoveTo(10,90);
LineTo(10,10,90,10);
PenThickness(.5);
MoveTo(10,20);
LineTo(12,20);
MoveTo(10,40);
LineTo(12,40);
MoveTo(10,60);
LineTo(12,60);
MoveTo(10,80);
LineTo(12,80);
MoveTo(20,10);
LineTo(20,12);
MoveTo(40,10);
LineTo(40,12);
MoveTo(60,10);
LineTo(60,12);
MoveTo(80,10);
LineTo(80,12);
// pie slices
PenColor(.5,1,.5);
FillColor(1,.5,1);
Pie(60,80,95,45,100,85,70,85,1);
FillColor(.5,.5,.5);
PenColor(0,.5,1);
Pie(60,95,95,60,100,100,70,100,0);
PenColor(0,0,0);
// re-trace the rim of a pie with a pen of different color
Arc(60,95,95,60,100,100,70,100);
// free-form color-filled polygon
PenColor(0,0,1);
FillColor(1,1,0);
```

Section 4. A gs tutorial

```
Polygon(50,30,90,30,60,15,70,40,80,15);
// ellipses and chords
Ellipse(15,50,30,90);
FillColor(1,0,0);
Chord(15,50,30,90,10,70,30,50,0);
// rectangle with rounded corners
PenColor(0,0,0);
PenThickness(.01);
FillColor(1,1,1);
RoundRect(50,50,80,55,3,5);
// markers and rounded dots
PenThickness(.2);
FillColor(1,1,0);
PenColor(0,0,1);
val square={-1,-1,1,-1,1,1,-1,1};
MoveTo(55,90);
Marker(square);
MoveTo(55,85);
Marker(square);
MoveTo(55,80);
Marker(square);
val star={-1,-1, 0,-.5, 1,-1, .5,0, 1,1, 0,.5, -1,1, -.5,0};
MoveTo(65,90);
Marker(star);
MoveTo(65,85);
Marker(star);
MoveTo(65,80);
Marker(star);
val star1={-1,-1, 0,-.5, 1,-1, .5,0, 1,1, 0,.5, -1,1, -.5,0, -1, -1};
MoveTo(70,90);
MarkerLine(star1);
MoveTo(70,85);
MarkerLine(star1);
MoveTo(70,80);
MarkerLine(star1);
PenThickness(2);
Point(60,90);
Point(60,85);
Point(60,80);
PenThickness(0);
MoveTo(58,92);
LineTo(62,92,62,78,58,78,58,92);
// text
```

```
FontName("Arial");
FontHeight(5);
MoveTo(10,110);
Text("Normal");
MoveTo(27,110);
FontBold();
Text("Bold");
MoveTo(40,110);
FontNormal();
FontItalics();
Text("Italics");
FontNormal();
FontInclination(45);
MoveTo(45,100);
Text("45 degrees");
FontName("Symbol");
MoveTo(65,110);
FontInclination(0);
PenColor(1,0,0);
Text("abcdefgh");
FontName("Arial");
FontHeight(1);
MoveTo(10,100);
Text("very small typefont");
Title("graph.txt");
```

If you are running Windows 95/98, some details may differ from the above picture.

SECTION 5. GRAPH WIZARD

Graph Wizard is a program which guides the user through a succession of steps to generate a scatter-plot. Once all the necessary data have been entered by the user, Graph Wizard generates an appropriate sequence of GS commands and sends it to Graph Server.

Before you start Graph Wizard you need to have your data available in a text file. In this file, numerical values can be written as integers or floating-point values. The data should be arranged as a matrix of rows and columns. Individual values must be separated by one or more spaces or tabs, but do not need to be padded with extra spaces in order to be visually aligned in vertical columns. Other requirements of the input data file are discussed below. Graph Wizard cannot read files in which data has been stored in non-text format (e.g., binary representation of data).

Section 5. Graph wizard

Each step in a series corresponds to a different dialog box in Graph Wizard. Before you can proceed to the next step, you must supply Graph Wizard with needed information. You can do so by filling the available text fields and other controls. In some cases, you can move back one or more steps without losing any of the data you already entered (you will be reminded whenever you are about to lose data by stepping backwards).

Graph Wizard displays help and hints at each step. Therefore, the following discussion will concentrate on the general meaning of each step, rather than describing the details of each operation. Each step is indicated by the title which appears in the corresponding dialog box.

The documentation on the CD contains illustrations of each of the following steps.

Graph Wizard – Introduction

This is an introductory screen which describes the scope of Graph Wizard, the characteristics it expects from the data contained in the input file, and the general way in which the program operates. This step requires no input. You can proceed to the next step by pressing the **Next** button.

Graph Wizard – Input file

Here you must enter the complete path (e.g., **C:\data\graph1.txt**) of the input file. As an alternative, you can use the **Browse** button to display the **Open Data File** dialog, which uses the "Explorer" interface familiar to Windows 95/98/NT users. You can use this dialog to navigate to an appropriate file, and subsequently double-click on its name or press the **Open** button. By default, this dialog displays files with **.TXT**, **.ASC** and **.DAT** names. You can choose **All files** in the **Files of type** list if you want to display also files with different extensions. Once you have the file path written in the **Input file** text box, press the **Next** button.

Graph Wizard – Select data

At this step you can select the data you intend to use. The purpose of this step is to eliminate any rows containing titles, comments, labels and any data which is not arranged as a matrix. At this stage, you **can** select also redundant rows of data (i.e., data which you do not intend to plot), as long as they are arranged in a matrix format. If your entire file satisfies the above criteria, you can select the whole file with the **Select All** button. Press the **Next** key when you are finished with this operation.

Graph Wizard – Edit the data to plot

This step has two separate purposes. In the **Data you selected from the input file** text box, you can inspect the data you selected during the previous step. Data items displayed in this box are separated by tabs. At the present stage, you can edit the data in this box in any way you wish, as long as adjacent data items are separated by one or more spaces or tabs. It is not necessary that the data items remain aligned in columns. It is necessary, on the other hand, that the number and ordering of data items per line remains consistent. The two radio buttons located below the text box allow you to choose if categories in your data are to be grouped by the vertical column or by the horizontal row.

Graph Wizard – Choose the data to plot

Here you have one more chance to inspect your data, displayed in a text box and grouped into rows or columns according to your earlier choice. You can no longer edit the data at this stage, but you can use the **Back** button if you need to do so.

This step also displays an empty text box under the heading **These are the defined data sets**. Before you can continue, you must select at least one data set from your data. Each data set will be displayed as a cluster of points in the scatter plot. All points belonging to a data set will be represented with the same symbol in the graph. Press **Add a new data set to the list** to continue. Once you are finished creating data sets (see the following step) press the **Next** button to proceed.

Graph Wizard – Add a new data set

This step is displayed as a new dialog box overlapping the preceding one. You can move the present dialog in order to see both. In this dialog, you will choose from which rows (or columns, depending on your earlier choice) to extract the x- and y-values to plot. You will also choose a marker, which will be displayed in the graph in correspondence of each data point belonging to this set.

You can choose among twenty types of geometric markers (which will be the same for all data points in a set), or an alphanumeric marker. The latter option needs some explanation. The "123" marker (as identified by its icon) will print a progressive number, starting from 1 and incrementing at each successive data point in the set. Similarly, the "ABC" and "abc" markers produce a succession of uppercase or lowercase characters, respectively, starting with "A" or "a". Upon reaching "Z" or "z", the sequence restarts from the beginning.

The "1" marker, on the other hand, assigns the same number (which corresponds to the number of the data set) to all data points in a set. Thus, the first set will have all data points marked as "1", the second as "2", and so on. Note that this applies also if other data sets use different markers. Thus, if the first data set uses a geometric symbol and the second the "1" marker, all points of the second set will be marked as "2". Similarly, the "A" and "a" markers yield an alphabetic identification of data sets.

Pressing the **OK** button will create the corresponding data set and return you to the preceding step, where you can create additional sets and/or delete previously-created ones. Pressing the **Cancel** button will return you to the preceding step without creating a data set.

Graph Wizard – General data settings

Here you can choose which co-ordinate range will be included in the graph. By default, Graph Wizard proposes a range sufficient to contain the data points and to allow for a "clean" margin all around the graph area. If you wish, you can choose a smaller or larger area. You can also choose the interval between tick marks placed along the axes. By default Graph Wizard chooses an interval appropriate to display 10 tick marks along the longest axis (the shorter axis receives proportionately fewer tick marks).

As a result of the above process, the tick marks may correspond to fractional values of measurement units. If you are planning to display the numerical values of measurement units alongside with the tick marks, you should probably change the spacing of the tick marks so that their associated numerical values are reasonable. For example, Graph Wizard may place the origin of the x-axis at co-ordinate 0.06 and place tick marks at co-ordinates 1.07, 1.902 and so on. If you wish to display the numerical values of the co-ordinates of tick-marks, it is reasonable to move the origin

Graph Wizard will not scale measurements differently along the x and y axes. If you want to "compress" a graph along the vertical or horizontal axis, you must do so by altering the input data.

Graph Wizard – General graph settings

These are settings that control the graphic appearance of the output. The first settings (area of the graph) control the co-ordinates of the plotting surface of the graph (excluding labels and text), expressed in GS plotting units. In most cases you do not need to alter these settings.

The next settings control whether the axes are drawn, and their thicknesses (also expressed in GS units). A thickness of 0.5 (the default) to 1 units is generally adequate.

Finally, you can choose whether tick marks are plotted along the axes, and their length and placement relative to the axes. The setting "inside axes" means that the tick marks will protrude into the plotting area, "outside axes" in the opposite direction.

Graph Wizard – Text settings

Here you may enter a text to be displayed along the axes (i.e., below the x-axis and to the left of the y-axis). The text along the y-axis is rotated 90° counter-clockwise relative to the horizontal axis. You can also choose the size of the text (i.e., its approximate height, expressed in GS units) and the font to use.

The dialog that allows you to choose a font displays all the fonts available on your system. This list may therefore vary from machine to machine. In addition to the font name, the dialog allows you to choose several other settings. *None of these additional settings has any effect in Graph Wizard.* They are displayed because Graph Wizard is using a font dialog that is part of the Windows operating system.

Graph Wizard – Finish

At this stage, you may choose whether to send the graph to Graph Server. As an alternative or additional option, you may save the graph to a file. The graph will be saved as a text file containing a sequence of GS statements.

You may additionally specify whether to write a log to a text file and/or display it in a window. The log contains a list of all GS statements issued to Graph Server and its corresponding replies. Possible replies include successful processing of the statement, detection of a syntax error, and explanatory messages returned by Graph Server. A log is not generated when the graph is sent only to a file.

SECTION 6. FREQUENTLY ASKED QUESTIONS

Other versions of Graph Server

Q – Is there a version of Graph Server for the Macintosh? Will a version of Graph Server for the Macintosh be released in the future?
A – No. There are no versions of Graph Server for computers, processors and operating systems, other than those specified in the section on hardware and software requirements. No new versions are planned.

Programming interface

Q – My compiler does not support calls to functions in DLL files. Can I still use Graph Server?
A – Yes (albeit in a slightly more awkward way). You can use the utility **SendFile.exe**. In the present version of GraphServer, you can also run Graph.exe from the command prompt, and provide a file name as an argument. This may change in future releases.

Section 6. frequently asked questions   101

Q – Graph Server complains about spurious characters being present at the end of each statement. Why?
A – Your program is probably appending a carriage return (or one or more other characters) at the end of each instruction, before sending it to Graph Server. Make sure that the last character of each statement is a semicolon.
   If your compiler insists in appending illegal characters at the end of an instruction, try adding an empty comment (i.e., the character pair //) at the end of each statement.

Q – I want to display a label containing the string "C:\Windows" in a graph. In order to make GS accept the backslash character, I wrote the string as "C:\\Windows" in my program. However, GS displays it as "C:Windows". What should I do?
A – You must take into account the escape sequences required by the language you are using to write your program, in addition to those required by GS. If you are writing your program in C or C++, you must enter each backslash as a double-backslash sequence. Each pair is turned into a single backslash by your compiler, and GS, in turn, will turn a backslash-pair into a single backslash. Therefore, you should write the string as "C:\\\\Windows" in the source code of your program.

Graphic output

Q – I am drawing a set of parallel lines spaced 5 units apart from each other. Pen thickness is set at 5 units. I expected the edges of adjacent lines to touch each other, but some of the lines are separated from each other by 1-pixel-wide gaps. While I resize the document window, 1-pixel gaps alternately appear and disappear between adjacent lines. How can I make sure that no gaps appear between adjacent lines?
A – If you must avoid gaps between adjacent lines, use the **Rectangle or Polygon** functions (with the appropriate fill colour and a pen thickness of 0) to draw thick lines, instead of the **LineTo function. Alternatively, use LineTo** with a line thickness sufficient to cause the edges of adjacent lines to overlap slightly.
   Gaps between adjacent lines are due to the fact that line thickness is expressed as a floating-point value in GS, but is approximated to an integer (which depends on several factors, including the sizes of the viewport and document window) when a graph is displayed. Consequently, line thickness can only be approximated in a display. This is especially evident in screen displays. As a rule, the sizes specified in the **Rectangle** function can be approximated in a better way than those specified in **PenThickness** and other functions.

Q – Thick lines are displayed with rounded ends and vertices. How can I draw a sharp vertex instead?

A – By default, GS rounds the ends and the convex vertices of lines. If you need sharp vertices, call **LineJoinMiter**. Call **LineJoinBevel** to display bevelled vertices (i.e., vertices with a triangular region of the tip cut-off). If you need square or flat line ends, call **LineEndSquare** or **LineEndFlat**.

Q – I want to draw a large colour-filled rectangle that encloses a number of smaller figures. However, when I draw the rectangle, it covers everything that was previously drawn in the same area. How can I avoid this?

A – When drawing a new graphic element, GS replaces everything that existed previously in the same area (or, if **Mode** is not set to default, performs logical operations between pre-existing and new elements). Therefore, you must plan your graph so elements that must lie in the "background" are drawn earlier, and elements in the "foreground" later.

Chapter 3

Methods for analysing a sample drawn from a single population

INTRODUCTION

All methods considered in this section are concerned with analysing a multivariate sample, drawn from a single population, from several conceptual standpoints. At the basic level of application, it is assumed that the data are multivariate normally distributed and interest is directed towards charting the geometrical and statistical properties of the sample. This is, however, not the only use that can be made of these methods and they can be profitably employed for seeking out the occurrence of heterogeneities in a data-set. This latter approach is often referred to as *exploratory data-analysis*. Another popular application is for reducing the dimensionality of a problem.

The subject of the multivariate analysis of single samples is an important one and there are several recent texts available on the subject. Jolliffe (1986) claims to be concerned with principal component analysis, but actually treats several other topics and his book could be categorized as presenting the application of the algebra of latent roots and vectors to the covariance matrices of single samples. Jackson (1991) is likewise directed towards the application of principal components, but also most other methods that treat a single multivariate sample find a place. Preisendorfer (1988) is mainly concerned with principal component applications in meteorology (with a specialized terminology) and is not directly accessible to geoscientists. Reyment and Jöreskog (1993) preferred to develop the subject in a wider context, with emphasis on identification of the model with respect to the concepts of **fixed mode** and **random mode**.

Fixed models dominate in geological work and, for that matter, in the Natural Sciences as a whole.

PRINCIPAL COMPONENT ANALYSIS

This method is probably the most commonly used one in applied multivariate statistics; it is, in many respects, also fundamental to the whole concept of multivariate statistics. Among its many applications, one of the most widely found is its use as a means of reducing the dimensionality of a problem without the loss of too much information: the new (and fewer) dimensions resulting from the principal component transformation are linear combinations of the observed variables. In its most elementary form, the method of principal component analysis consists of the extraction of the latent roots and vectors of either the covariance matrix or the correlation matrix. The treatise by Jackson (1991) is recommended for all who wish to gain a deeper insight into the many ramifications of the method, notwithstanding that it roams off into all sorts of other fields and is regrettably deficient in its treatment of developments in applications in the Natural Sciences. Principal component analysis is a so-called R-mode procedure. We have used the term "latent" above with respect to the roots and vectors of a matrix. Other terms used for latent roots and vectors are eigenvalues and eigenvectors, proper values and vectors, and characteristic roots and vectors, which latter amalgamation may just have priority over the designation chosen by us. *Eigenwert* is the German translation of characteristic root, which, in a period of anglophonic linguistic decline, was back-translated as the truly horrible hybrid "eigenvalue", widespread in North American publications, and, lamentably, beginning to infect British literature. *Valeur propre* is the French rendition of latent root, which also wandered back home as "proper value", which is quite silly in English, but sensible, of course, in French. The algebra of latent roots and vectors is rather complicated. If you wish to learn the basic principles, Reyment and Jöreskog (1993) have a section on the subject. Of the more technical accounts, Bellman (1960) and Gantmacher (1965) are mines of information.

The application of latent roots and vectors in physics, astronomy, engineering and chemistry is widespread and the step to statistics was not all that of a big one to take some 60 years ago. Closely related computationally (though not conceptually) to principal component analysis is *Factor Analysis*. The subject of Factor Analysis in the Natural Sciences has recently been treated by Reyment and Jöreskog (1993). It is considered only briefly in a later section.

The basic features of principal component analysis (but not the theoretical derivation thereof) depend on the extraction of the latent roots and vectors of a square symmetric matrix. Consider a real, square symmetric matrix, **R** (the correlation matrix, for example). A latent vector **u** of **R** is given by the relationship

$$\mathbf{Ru} = \mathbf{u}\lambda \qquad (3:1)$$

where λ is a scalar, called the latent root, to be estimated. Equation (3:1) can also be written, by rearrangement, as

$$(\mathbf{R} - \lambda \mathbf{I})\mathbf{u} = \mathbf{0} \tag{3:2}$$

where **I** is the *identity matrix*, a matrix with ones down its diagonal and zeros in all other positions, and **0** is the *null vector*, a vector with all its components equal to zero.

The first step in finding **u** and λ is to solve the determinantal equation

$$|\mathbf{R} - \lambda \mathbf{I}| = \mathbf{0} \tag{3:3}$$

This expands to a polynomial with as many roots as there are dimensions for a full-ranking matrix **R**. The lambdas of (3:3) enter into a diagonal matrix where they constitute the diagonal elements, all other positions being zero.

Thus,

$$\Lambda = \begin{pmatrix} \lambda_1 & 0 & 0 & \ldots & 0 \\ 0 & \lambda_2 & 0 & \ldots & \\ \ldots & \ldots & \ldots & & \\ 0 & 0 & 0 & \ldots & \lambda_p \end{pmatrix}. \tag{3:4}$$

with the lambdas ranged in descending order of magnitude. That is

$$\lambda_1 > \lambda_2 > \ldots > \lambda_p. \tag{3:5}$$

By the same token, the latent vectors corresponding to the lambdas are found by solving the simultaneous equations (cf. 3:2)

$$(\mathbf{R} - \lambda \mathbf{I})\mathbf{u} = \mathbf{0}$$

for each lambda in turn. These vectors, when standardized to have unit length, can be grouped into a matrix **U**. This is an interesting, and useful, matrix because its columns are mutually orthogonal. In statistical terms, the latent vectors are linear combinations of the original p variables that are uncorrelated with each other. It comes sometimes as a surprise that the components of a latent vector are all negative, or that different programs return latent vectors with 'mirrored' signs on their components. This is an expression of the fact that the sign of a latent vector is indeterminate. Summing up what has been done, we have that in the most general form,

$$\mathbf{RU} = \mathbf{U}\Lambda \tag{3:6}$$

which, by rearrangement of terms, is equivalent to

$$\mathbf{R} = \mathbf{U}\Lambda\mathbf{U}^\mathrm{T} \tag{3:7}$$

Let us ponder over this result for a moment. There are four important things to notice.

1 Equation (3:7) shows that a square symmetric matrix can be decomposed into two equal orthogonal matrices and a diagonal matrix. The superscript "T" attached to the second **U** indicates that it is transposed – it is lying on its side, as it were. An equivalent, and more widely used, symbol is a dash placed after the symbol for the matrix to be transposed which we have avoided in the present text, owing to the ease with which it can be overlooked on a word-processor.

2 The non-zero elements of Λ lie along a diagonal. The sum of these diagonal elements is exactly equal to the sum of the diagonal elements of **R**, which is a most valuable property. For one, it shows that the sum of the diagonal elements of Λ is equal to the sum of the variances of the original matrix. This sum is known as the *trace* or *spur* (= German for 'trace') of a matrix. The relationship implied is written:

$$\mathrm{tr}\Lambda = \mathrm{tr}\mathbf{R}$$

3 Each diagonal element of **R** has a corresponding vector, a linear combination of the original variables. Hence, the *rotation* engendered in the principal component transformation makes p new uncorrelated variables, linear combinations of the p original, correlated ones. This is again a valuable property and one that can be used in many connexions in multivariate statistical analyses. Nothing has been altered, added or changed. All we have done is that we have transformed, by the rotation of axes, a set of correlated variables into an equivalent, new set of uncorrelated variables. The practical significance of this manipulation is far-reaching in many aspects of statistical work in which data-reduction is a prime concern.

4 The product of the diagonal elements of Λ yield the numerical value of the matrix, its *determinant*. This is still another useful attribute in that this determinant is the same as that obtained from finding $|\mathbf{R}|$ directly. If you already have the latent roots of **R**, it becomes a trivial matter to compute its determinant, a task that is onerous for larger matrices by the classical method. For more details, we refer you to Chapters 2 and 3 in Reyment and Jöreskog (1993).

5 There is a further point we shall briefly mention now and return to later; it is a very critical one. The *rank* of a square symmetric matrix is equal to the number of non-zero latent roots of that matrix. This is of consequence when dealing with compositional data. (Note, the mathematical use of the word "rank" is not the same as that in the "computerese" of FORTRAN-90 programming.)

6 Although the correlation matrix is the popular choice in principal component analysis, it is necessary to point out that there are fewer significance tests available for correlations than for principal components computed from covariances. It is for this reason that we advocate the use of the logarithmic transformation of the data wherever possible and reasonable. The logarithmically transformed data

Principal component analysis

yield a result that differs little from that produced by standardized observations and there is the added advantage that all appropriate tests are available. In essence, the steps outlined above form the basis of the programs relevant to this chapter. Figures are often more persuasive than symbols which is why we shall now look at a simple example for four variables. You can run it just by typing *pcomp1* at the *DOS* prompt. The salient features of the results of the calculations are displayed in **Box 3**. The matrix to be decomposed by *pcomp1* is the quadrivariate covariance matrix supplied as the first array in **Box 3**. The data are located in a file called *pcomp.dat*.

Box 3: A simple example of the extraction of latent roots and vectors in statistics

Program: **pcomp1**

Data: **pcomp.dat**

	var 1	var 2	var 3	var 4
var 1	**0.26643**	0.08518	0.18290	0.05578
var 2	0.08518	**0.09847**	0.08265	0.04120
var 3	0.18300	0.08265	**0.22082**	0.07310
var 4	0.05578	0.04120	0.07310	**0.03911**

This starting matrix, which we shall call **R**, is square and it is symmetric. The diagonal elements, printed in bold type, are the (univariate) variances of the four variables in turn. The output takes the form annotated below.

The program begins by identifying the number of variables in the exercise. (N.B. the number of decimals is an artefact of the program and not meant to imply, necessarily, that level of accuracy in the calculations.)

dimensionality of the problem = 4

simple principal component analysis for covariance matrix

latent roots = principal component variances (the lambdas of the main text):

0.48788   0.07238   0.05478   0.00979

latent vectors = principal component loadings ($\mathbf{U}^T$):

component 1
0.68672   0.30535   0.62366   0.21498

component 2
−0.66908   0.56748   0.34333   0.33530

component 3
−0.26510   −0.72961   0.62717   0.06367

component 4
0.10228   −0.22892   −0.31597   0.91504

The latent roots are the lambdas found by solving eqn. (3:3). The latent vectors are the **u**'s obtained from eqn. (3:2), but with something extra done to them. The vectors, as computed, must be constrained so as to make them compatible from sample to sample since there is no unique solution to (3:2). This implies that if **u** is a solution, so is c**u** a solution, c being any scalar, and so on. The way out of this awkward situation is to make the vectors have unit length and this has been done here (it is sometimes referred to as *normalizing* the vectors). Hence, the four latent roots are the diagonal elements of Λ and the four rows of latent vectors are the *columns* of **U**. We have output the matrix in this form to show you what the transpose of **U** looks like.

If you add up the diagonal elements of **R**, and the four lambdas, you will get the sum 0.62482 in both cases. Another thing you can check is the sum of the squares of the components of the latent vectors. You should get a one, within rounding limits, in each case. A second instructive exercise is to multiply and add corresponding components in any two vectors and then take the cosine of this. The result should be 90°, within the limits of rounding, thus proving that the vectors are at right angles to each other, i.e. orthogonal, and hence, uncorrelated with each other.

Scaling the latent vectors

We have called the latent vectors in the foregoing examples principal components. This is a common enough usage, but in many applications, the principal components are taken to be the latent vectors scaled by multiplication with the square root of the

Principal component analysis

corresponding latent roots. This amounts to scaling each latent vector so that its squared length equals the matching latent root. This manipulation in no wise changes the sense of direction of the vectors, since all vectorial elements retain the same proportionality. If $\hat{\mathbf{A}}$ denotes the scaled latent vectors, the appropriate formula for producing these is:

$$\hat{\mathbf{A}} = \mathbf{U}\Lambda^{1/2} \tag{3:8}$$

What this step implies can be shown by invoking the program *pcaident* in the following manner.

C:\pcaident‹matinv2.dat

Look now at the listing from this presented in **Box 4**.

Box 4: *Illustrating the concept of scaling of principal components*

Program: **pcaident**
Data: **matinv2.dat**

(N.B. The five decimals in the results are not meant to imply that level of accuracy, but are an artefact of the computations made by the program.)

The input:

The covariance matrix

$$\begin{bmatrix} 3.268 & -0.912 & -0.645 & -1.464 \\ -0.912 & 3.211 & -1.160 & -0.914 \\ -0.645 & -1.160 & 2.403 & -0.169 \\ -1.464 & -0.914 & -0.169 & 2.916 \end{bmatrix}$$

The latent roots

4.64516   4.26062   2.57664   0.31558

The latent vectors

$$\begin{bmatrix} & 1 & 2 & 3 & 4 \\ -0.69800 & 0.40113 & 0.30349 & 0.50969 \\ -0.16537 & -0.83253 & -0.13739 & 0.51056 \\ 0.23696 & 0.37311 & -0.75683 & 0.48151 \\ 0.65521 & 0.08227 & 0.56234 & 0.49769 \end{bmatrix}$$

The principal components (vectors scaled by square-root of lambda) cf. eqn. (3:8).

$$\begin{bmatrix} & 1 & 2 & 3 & 4 \\ -1.50437 & 0.82799 & 0.48716 & 0.28633 \\ -0.35641 & -1.71845 & -0.22053 & 0.28681 \\ 0.51071 & 0.77015 & -1.21485 & 0.27049 \\ 1.41216 & 0.16981 & 0.90267 & 0.27959 \end{bmatrix}$$

Principal components (vectors scaled by inverse square-root of lambda) cf. eqn. (3:9).

$$\begin{bmatrix} & 1 & 2 & 3 & 4 \\ -0.32386 & 0.19434 & 0.18907 & 0.90730 \\ -0.07673 & -0.40333 & -0.08559 & 0.90885 \\ 0.10995 & 0.18076 & -0.47149 & 0.85713 \\ 0.30401 & 0.03986 & 0.35033 & 0.88595 \end{bmatrix}$$

Further notes on scaling of latent vectors

It is appropriate at this point to consider some further properties of square symmetric matrices insofar as they apply to the scaling of latent roots and vectors. In principle, there are three ways of scaling the vectors, of which we have already learned two. To summarize, these are:

1   The vectors are scaled to unity. That is nothing is done to them other than to normalize them. These are the "raw" latent vectors obtained from the covariance or correlation matrix as in eqn. (3:7).

2 The most commonly used scaling is the one we have just considered in this section, that is, the one that produces the results displayed in **Box 4** and which is expressed by the multiplication of eqn. (3:8).

3 There is a third method of scaling that is seldom found in applications in biology and geology, but which is widely used in quality control in industrial connexions (cf. Jackson, 1991). This is the scaling produced by *dividing* the elements of the latent vectors by the square root of the corresponding latent root.

$$\mathbf{W} = \mathbf{U}\Lambda^{-\frac{1}{2}} \qquad (3:9)$$

This method is also illustrated in **Box 4**.

These three scaling relationships bring to light some interesting identities of practical significance in statistical work, not least with respect to programming of computational procedures.

The product

$$\mathbf{A}^T\mathbf{A} = \Lambda$$

which is the diagonal matrix of latent roots, where $\mathbf{A}$ is as defined in eqn. (3:8).

The product

$$\mathbf{A}\mathbf{A}^T = \mathbf{S}$$

returns the covariance matrix $\mathbf{S}$ with which we started out.

The product

$$\mathbf{W}^T\mathbf{W} = \Lambda^{-1}$$

which is the diagonal matrix of reciprocal latent roots. Here, $\mathbf{W}$ is as defined in eqn. (3:9).

The product

$$\mathbf{W}\mathbf{W}^T = \mathbf{S}^{-1} \qquad (3:10)$$

produces the inverse of the matrix with which we started. This latter identity can be put to good use in some programming situations requiring a (not too accurate) matrix inversion, and in particular, the Moore–Penrose generalized inverse, which achieves prominence in the analysis of compositional data in situations in which singular matrices must be inverted (i.e. a square symmetric matrix with a zero determinant). For convenience, this inverse is now presented. This pseudo-inverse of the centred log-ratio covariance matrix by the Moore–Penrose algorithm is defined as in eqn. (3:11):

$$\Gamma^- = \lambda_1^{-1}\mathbf{a}_1\mathbf{a}_1^T + \ldots + \lambda_d^{-1}\mathbf{a}_d\mathbf{a}_d^T \tag{3:11}$$

The *d*-th term, $\mathbf{a}_d\mathbf{a}_d^T$ is zero owing to the *singularity* of the log-ratio covariance matrix.

If you use the inverse matrix as input to **pcaident.exe** you should obtain the following result on comparing the output obtained by analysing S and its reciprocal, S^{-1}:

(a) The latent roots are different in that the diagonal of one matrix is exactly the reciprocal of the other.
(b) The latent vectors for both matrices are the same.
(c) The *z*-score vectors for one become the *y*-score vectors for the other, and vice versa.

z-scores and *y*-scores

The second method of scaling, the one that leads to *z*-scores, and which is the type used in this book, leads to the scores defined by the relationship:

$$z_i = \mathbf{u}_i^T[\mathbf{x} - \bar{\mathbf{x}}] \tag{3:12}$$

where $\mathbf{x}$ denotes any of the multivariate observations in the sample, $\bar{\mathbf{x}}$ is the mean vector of the sample and the $\mathbf{u}_i$ are the latent vectors. The principal components on *z*-scores have a practical advantage in geochemistry, for example, inasmuch as they are in the same units as the original variables. That is, grams per litre remain grams per litre.

The *variances* of the scores are the squares of the corresponding latent roots, because by eqn. (3:8)

$$\mathbf{A}^T\mathbf{S}\mathbf{A} = \Lambda^2$$

The scores produced by the third type of scaling are computed as

$$y_i = \mathbf{w}_i^T[\mathbf{x} - \bar{\mathbf{x}}]. \tag{3:13}$$

The attractive feature of these scores for quality control (which, of course, includes ores) is that the variance is

$$\mathbf{W}^T\mathbf{S}\mathbf{W} = \mathbf{I}$$

the identity matrix. Hence, all variances are unity.

As demonstrated by the program **pcaident.exe**, there is a simple relationship between the two kinds of scores, to wit:

$$y_i = \frac{z_i}{\sqrt{l_i}}$$

and

$$z_i = \sqrt{l_i}\, y_i$$

PRINCIPAL COMPONENT FACTOR ANALYSIS

For many years now, it has been customary to rotate the scores obtained from a standard principal component analysis of the correlation matrix by some appropriate technique, albeit with the benefit of some minor adjustments. The idea comes from the realm of early psychometry, but has now been largely supplanted by more adequate models. This is usually known as "factor analysis" in geological and biological work, although it is more correctly designated as *principal component factor analysis*, or principal component analysis with rotation of the axes to some kind of simple structure (Reyment and Jöreskog, 1993). There is a considerable theoretical difference between the aims and methods of true factor analysis, as appropriate in psychometrical work, and almost all applications occurring under that name in the natural sciences. An example of true factor analysis in oceanology is given in Reyment and Jöreskog (1993) (see the Ivorian oceanographical study).

In general terms, the technique of principal component factor analysis can be said to be concerned with sample quantities and there is no attempt at trying to estimate the population counterparts, such as pertains in true factor analysis in psychometry. Hence, the results obtained can only be interpreted at the level of the sample on which the calculations were performed. This is the **fixed mode model** as opposed to the **random mode model** (Reyment and Jöreskog, 1993, Chapter 4, section 4.2). The principle of fixed mode multivariate analysis was introduced earlier on (p. 17) in relation to the "constant weight stratagem").

The procedure of rotating a principal component solution has in orthodox statistical spheres long been regarded as the ultimate manifestation of charlatanism. However, over the last decade, many professional statisticians have seen the usefulness in doing this in some practical connections and it is no longer considered poor form to rotate the axes of a principal component analysis in the search for analytical enlightenment (Seber, 1984; Jackson, 1991; Preisendorfer, 1988). The program *pcomp2.exe*, now to be introduced, allows the varimax rotation of axes as an option.

You should be aware of one thing at least if you plan on rotating the principal component axes. The uniqueness of the latent vectors is achieved by the descending order of importance of the associated latent vectors. Rotation of axes upsets the variance-maximization criteria of principal component analysis.

AN EXPANDED PROGRAM FOR PRINCIPAL COMPONENTS

The program *pcomp2.exe* provides a more complete principal component analysis than either of the two routines with which you have already been familiarized. The scores output by the program are placed in files for subsequent graphical examination in *Graph Server*. An example of the graphical output of the scores for the principal components is shown in Fig. 27(a). Note that this plot yields unequivocal evidence of heterogeneity in the data, including three strongly atypical values. Fig. 27(b) displays the rotated principal component factor scores for the same set of data and for the same latent vectors. There is little difference in the two figures

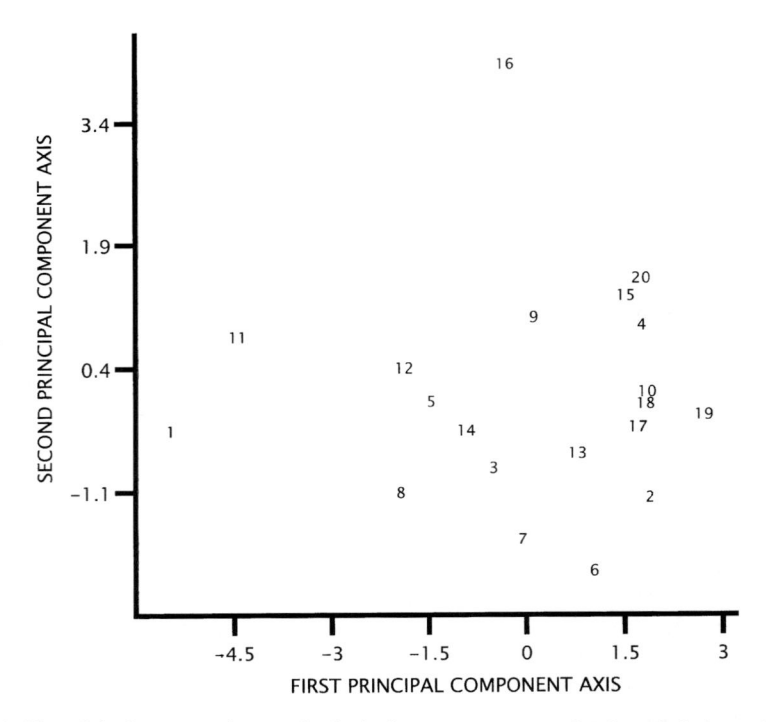

Fig. 27(a). Plot of the first two columns of principal component scores for the *Afrobolivina* foraminiferal data. Three points plot at a considerable distance from the main cluster. These are for observations on megalospheric specimens of the species.

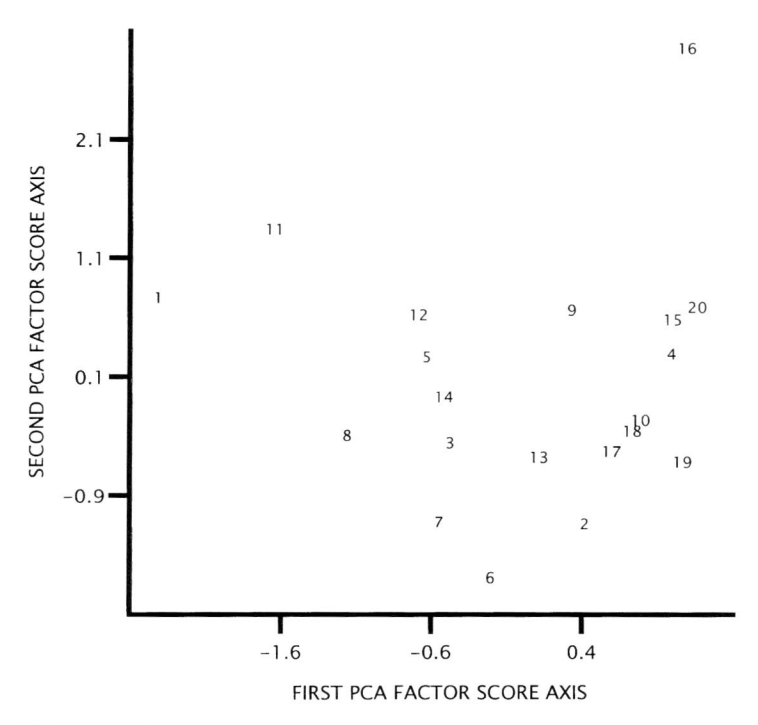

Fig. 27(b). Plot of the first two columns of the varimax rotated principal component factor scores for the *Afrobolivina* foraminiferal data. This graph does not differ in any essential manner from Fig. 27(a).

and we conclude that rotation of the axes was a needless manoeuvre. It can also be added here that a reification of the components of the latent vectors would be quite unjustifiable.

Principal component analysis is the most universally available of all multivariate methods, the reason for this being that it provides a satisfactory way of describing the *geometrical* configuration of a cluster of points. In tri-dimensional space, a cluster of perfectly multivariate-normally distributed points will appear as an ellipsoid something like a rugby football hovering in space. The principal components are, in simple geometrical terms, therefore no more than the principal axes of this ellipsoidal body (hence the origin of the name of the method). The first, and largest, component is just the major axis of the ellipsoid. The next longest, the second principal component, is the second principal axis (or largest of the minor axes), and the third is the shortest of the three axes. The principal component coefficients, the latent vectors, are the algebraic specifications of the principal axes of the ellipsoid, the direction cosines of these lines. The length of each axis is proportional to the corresponding latent root. So much for the solid geometry, but how can this be transformed into statistical terms?

1   The location of the latent vectors along the principal axes of the hyperellipsoids (the appropriate geometrical term when $p > 3$) in effect positions them so as to coincide with the directions of greatest variance. The latent vector associated with the second greatest latent root locates the direction of maximum variance at right angles to the first, and so on.

2   The location of latent vectors can be interpreted as a rotation or transformation. The original axes, which are given in terms of the original, correlated variables, are rotated to new positions, and thus to a "new set" of uncorrelated variables, which account for the variance of the data in decreasing order of importance (i.e. expressed by the magnitudes of the latent roots). It can often prove useful to be able to reduce the dimensionality of a problem to a few "new" variables, particularly in such geological connections as superimposing trends on maps, borehole analysis, and in palaeoecology.

3   A latent root of zero indicates that the corresponding minor axis is of zero length which, in turn, is good evidence for assuming that the rank of the data matrix is less than its dimensionality. This is a situation we have met in the case of constrained variables.

Instructions for using the program **pcomp2**

Line 1: The dimensionality of the problem.

Line 2: The title of the job

Line 3: 1 Size of the sample
        2 1 for covariances, 0 for correlations
        3 1 for logarithms, 0 for raw data
        4 1 for principal component scores, else 0
        5 1 for correlations between the principal components and the original variables, else 0
        6 1 for a principal component factor analysis

Line 4: The data-matrix in free format.

The trial data are in the file *afrobol.dat*, being measurements on seven dimensions observed on the test of the Campano-Maastrichtian bolivinid foraminifer *Afrobolivina afra* from the subsurface of coastal Nigeria. They are length of the test, maximum breadth of the test, heights and breadths of the last two chambers and the diameter of the proloculus. The first five of these variables appear in the analysis summarized in **Box 5.** The principal component scores are stored in the file **pcascore** If you requested a principal component factor analysis, the pertinent

An expanded program for principal components

scores are located in the file **facscore**. These files are for subsequent use with *Graph Server*. A simple step-by-step account of how to use *Graph Server* in its most unsophisticated mode is given at the end of this chapter (on p. 153).

The more important features of the work done by the program are now summarized in **Box 5**. We have asked for a principal component factor analysis and for the calculations to be made on the correlation matrix.

Box 5: Principal component analysis of foraminiferal data: part of the program output.

Program: **pcomp2**
Data: **afrobol.dat**

Principal component factor analysis requested

Principal components for correlations

THE AFROBOLIVINA AFRA DATA

Number of dimensions = 5

Sample-size = 70

vector of means

73.516   31.306   25.241   15.980   14.543

standard deviations for each variable

21.167   4.542   4.182   3.047   3.169

Covariance matrix

	1	2	3	4	5
1	448.0267	65.3602	72.4797	31.3497	37.8573
2	65.3602	20.6253	15.5819	7.8943	9.9415
3	72.4797	15.5819	17.4868	7.8499	9.5386
4	31.3497	7.8943	7.8499	9.2863	4.0687
5	37.8573	9.9415	9.5386	4.0687	10.0431

trace of covariance matrix = 505.4682

Correlation matrix

	1	2	3	4	5
1	1.0000	0.6799	0.8189	0.4860	0.5644
2	0.6799	1.0000	0.8205	0.5704	0.6907
3	0.8189	0.8205	1.0000	0.6160	0.7198
4	0.4860	0.5704	0.6160	1.0000	0.4213
5	0.5644	0.6907	0.7198	0.4213	1.0000

Correlations used for computations

Latent roots

3.58375   0.60007   0.43854   0.25816   0.11948

percentages of the total variance represented by the latent roots

71.67   12.00   8.77   5.16   2.39

latent vectors by columns

	1	2	3	4	5
1	0.4476	0.1431	−0.7153	0.3113	0.4131
2	0.4741	0.0840	0.1108	−0.8260	0.2715
3	0.5010	0.0677	−0.1647	−0.0060	−0.8469
4	0.3777	−0.8739	0.2097	0.1927	0.1115
5	0.4257	0.4518	0.6364	0.4286	0.1612

Latent vector times the square root of the latent roots

	1	2	3	4	5
1	0.8473	0.1109	−0.4737	0.1582	0.1428
2	0.8975	0.0651	0.0733	−0.4197	0.0938
3	0.9485	0.0525	−0.1090	−0.0030	−0.2927
4	0.7151	−0.6770	0.1389	0.0979	0.0385
5	0.8058	0.3500	0.4214	0.2178	0.0557

Scores for latent vector times square root of latent root

1	−3.0307	0.6170	−1.1292	−0.2096	0.3218
2	3.6989	−0.7273	−0.5064	0.1791	0.1549
3	1.1683	−0.2103	−0.5978	0.0416	−0.1877
4	3.0721	−0.3349	−0.0784	−0.4145	0.1582

....

67	−0.0771	−0.6995	0.0024	−0.2038	−0.1066
68	−0.9388	−1.1072	−0.0608	0.1525	−0.0338
69	−1.4781	−0.4973	−0.0421	0.1947	0.0252
70	0.0776	0.2862	−0.1702	0.4889	0.0571

Test for isotropy of the last $m - 1$ latent roots of the correlation matrix (i.e. equality of the last $m - 1$ latent roots).

Chi-square = 37.81 for 9 degrees of freedom. The hypothesis of isometry is rejected.

Principal component factor analysis by varimax rotation

Varimax factor matrix

Var	Comm.	1	2
1	0.7302	0.7859	−0.3355
2	0.8097	0.8058	−0.4005
3	0.9024	0.8433	−0.4373
4	0.9696	0.2713	−0.9466
5	0.7719	0.8718	−0.1086
variance		56.24	27.44
Cum. Var		56.24	83.68

Varimax factor score matrix

1	−0.2051	1.3155	−1.2534	−1.6342	−1.4335
2	0.2721	−1.5685	0.6597	1.5161	0.1443
3	0.1024	−0.4675	0.8988	0.2363	−0.4868
4	0.4542	−0.9165	0.4205	0.8598	0.1443

....

67	−0.6114	−0.9927	0.1814	0.6629	−0.6446
68	−1.1640	−1.4554	−0.2490	1.0895	−0.8024
69	−0.7766	−0.5037	−0.4882	0.3347	0.4868
70	0.2613	0.3996	−0.0577	−0.1575	0.4598

The plot of the first two columns of the matrix of *principal component scores*, using Graph Server is illustrated in Fig. 27(a).

Zero latent roots

What do we do with latent roots that are zero (for reasons other than, for example, the closure constraint), or almost zero? Should they be reified? The answer is yes because such roots are connected with invariant linear relationships in the original variables. The significance of such a reification in quantitative petrology is that the invariant association can disclose the existence of a ratio relationship between parts such as are much favoured by petrologists; there is an example in Reyment (1978b). The implication is, for z-scores for example, that

$$z_i = \mathbf{u}_i^T[\mathbf{x} - \bar{\mathbf{x}}] = 0 \text{ for any } \mathbf{x}.$$

PRINCIPAL COMPONENTS AND CROSS VALIDATION

One of the aims of principal component analysis is to achieve a "parsimonious description" of a multivariate data-set. There is, therefore, a decision required as to how many principal components are to be retained in any given situation. There is no hard and fast rule for this, although at least one "rule-of-thumb" exists. Compute the cumulative percentage variance contribution for successive values of the number of latent roots extracted. The appropriate level at which to stop extracting roots is usually taken at the point at which 95% of the trace of the covariance (correlation) matrix has been accumulated (Reyment and Jöreskog, 1993, p. 98). Another, less popular, rule is to retain all latent vectors that are at least as variable as the original variables, namely, equal to or greater than 1 for standardized variables (correlation matrix). There is also the "screes" method which indicates a cut-off for significant roots at a point where the graph of latent root against order falls off flatly (like the screes of débris on a mountain slope).

Applications of principal component analysis in chemometrics have shown that an ad hoc technique known as **cross-validation** can prove useful for obtaining answers to such practical questions as:

(a) How many principal components should be retained in an analysis?
(b) How many, if any, variables can be considered to be redundant (and could eventually be excluded from subsequent studies)?

(c) Do any of the specimens in the sample deviate statistically from the others?

You will probably recognize the scope of cross-validation as being sample-oriented. Cross-validatory analysis is an exploratory technique that looks for interesting relationships in the sample. It is often found to be most useful as a first step towards making a complete multivariate analysis. Analytical chemists seem to be completely convinced of the worthwhile nature of cross-validation in their work and it is therefore surprising, not to say alarming, that inorganic geochemists have not adopted the technique to any extent as far as we are aware. The literature on organic geochemistry (including geochemistry) does, on the other hand, contain many applications of cross validation; unfortunately, the authors of these applications seem to be unaware of the special requirements imposed by compositional data such as are of wide occurrence in chemical data.

Krzanowski (1987a,b) produced a synthesis of methods for obtaining an answer to the questions enumerated above, as well as other, more complicated ones and it is his combination of techniques that we employ here. The present concept of cross-validation derives from a paper by Lachenbruch and Mickey (1968) in which a technique designated as 'leave-one-out' was promulgated for improving a discriminant analysis. There are other ways of going about a comprehensive cross-validation analysis, as is shown in the pages of the journal *Chemometrics*, but in all essential aspects, these do not differ from the procedure used here.

We introduce the topic via the program furnished for doing the calculations. The program *pcvalid.exe* performs the necessary computations and is a Fortran 90 restructuring of Professor Wojtek Krzanowski's original pilot program in FORTRAN IV (written in 1987). This is a "computer-intensive" technique that makes use of the iterative procedure known technically as "jackknifing", one of John Tukey's folk-humouristic fanciful terms (the likeness comes from comparing the "opening" and "closing" of the data-set to remove and replace an observational vector to the opening and shutting of a patent knife). A related computer-intensive technique was given the equally whimsical name "bootstrapping" by D. Efrom, the likeness being based on comparing the computational exertion required with the expression "to pull oneself up by one's bootstraps (= shoelaces)".

Instructions for using the program **pcvalid**

Line 1: The title of the job.

Line 2: In free format, the following;
    1 sample size

2   number of variables
  3   0 if the analysis is to be made on the covariances or 1 if the correlation matrix is to be employed.

Line 3: Order in which the variables are to appear in the analysis. This will usually be in the "natural" order, but some other ordering might be of interest.

Line 4: Data-matrix in free format

Output Details

The output encompasses the following features:

  1   An evaluation of the role of each variable. Variables contributing little to the analysis can be deleted in a new round of calculations.
  2   The effect of successively deleting observations can be ascertained. This step discloses the presence of influential and, or, atypical observations that might otherwise have been missed in a routine appraisal of the data. The columns headed by positive integers are largely dominated by variances, the columns headed by negative integers are more the domain of the covariances.
  3   The third special result yielded by the program concerns the logical number of principal components to retain.

There is a discussion of the technique, with an oceanological example, in Reyment and Jöreskog (1993, pp. 115–121). A geochemical example (analyses of alkaline rocks) will serve to introduce the main features of the synthesis of techniques. The data were selected from a compilation in treatise format made by Sørensen (1974) for alkaline eruptive rocks, the details for which are given in **Box 6**. Note, that we have gone ahead of the next section in that these data are constrained to have row-sums of 100%, for which reason an appropriate log-ratio transformation of the data-matrix has been used. Note also, on this occasion we have used the latent roots and vectors of the correlations – remember that these data are now in the form of ratios, not raw observations. Look now at **Box 6.**

You can do the analysis yourself by typing at the DOS prompt

C:*pcvalid*‹*alkaval.dat*

Box 6: Constrained cross-validational log-contrast principal component analysis for the chemical composition of alkaline rocks (Sørensen, 1974).

Program: **pcvalid**

Data: **alkaval.dat**

******** KRZANOWSKIS cross-validation principal components

(N.B. The following text is part of the print-out from the program; it is not meant to imply that the number of decimals listed reflects the actual accuracy of the program.)

The log-ratio data for Canary Islands, southern South Atlantic, Pacific, northern South Atlantic
All 12 parts (oxides) are included in the analysis:

1	2	3	4	5	6	7	8	9	10	11	12
Si	Ti	Al	Fe^{+++}	Fe^{++}	Mn	Mg	Ca	Na	K	H_2O	P

PCA by JACK-KNIFING procedure

Number of parts = 12

No of samples = 59

Computed for the correlation matrix

Latent roots of correlation matrix

6.5918 1.4132 1.0640 1.0107 0.7818 0.3664 0.2420 0.2002 0.1541 0.1092 0.0667 0.0

Here the 12th latent root is nought because of the compositional constraint

Principal components
(These are the latent vectors, listed vector by vector.)

Component 1
−0.32840 0.30261 −0.31562 0.07515 0.25594 −0.20567 0.35618 0.34939 −0.34781 −0.35357 −0.17861 0.24634

Component 2
0.29777 −0.03447 0.21027 0.18956 0.50628 0.41518 −0.04966 0.19759 0.09642 −0.10350 −0.56151 −0.15649

Component 3
−0.08963 −0.23213 −0.21756 0.91252 −0.01724 −0.16889 −0.08483 −0.04143 0.08858 0.07939 0.06041 −0.05032

Component 4
0.20963 −0.04417 0.03676 −0.02856 0.28406 −0.18673 0.16280 0.23439 −0.17252 −0.11761 0.48792 −0.68795

Component 5
−0.07725 −0.50162 −0.33378 −0.06351 0.09750 0.63456 0.14103 −0.09543 −0.22147 −0.14883 0.30029 0.15630

Component 6
0.02377 −0.54230 0.30444 −0.12628 0.37817 −0.46163 −0.12193 0.00502 0.10678 −0.23551 0.08255 0.38991

Component 7
−0.47137 0.17943 0.48885 0.14828 −0.04150 0.27318 −0.45422 0.25750 −0.13237 −0.20268 0.27706 0.01811

Component 8
−0.31764 0.08008 −0.43515 −0.21315 0.49085 −0.02817 −0.41840 0.07095 0.24015 0.41237 0.05992 −0.06907

Component 9
0.07582 −0.04530 0.19403 0.05682 0.15424 −0.07401 −0.06989 −0.13854 −0.76787 0.54075 −0.08174 0.09619

Component 10
0.34486 0.44378 −0.12270 0.10770 0.26726 0.01287 −0.27525 −0.54561 −0.08956 −0.35036 0.23131 0.16780

Component 11
−0.52158 0.01662 0.27157 0.01706 0.24966 0.02124 0.40248 −0.58938 0.11828 0.00679 −0.08488 −0.24547

Effect of deleting each variable in turn (with subsequent replacement)
Residual sums of squares = Procrustean fit of new scores to old scores
(Schönemann and Carroll, 1970)

Sizes of principal component-spaces being compared

	Principal Components				
Variable removed	1	2	3	4	5
1	3.7433	8.6082	11.6615	5.7798	5.0430
2	3.3670	3.4288	9.1491	6.6865	11.6057
3	3.5357	5.9242	14.7361	7.5780	7.1046
4	0.3301	2.7166	117.3480	98.9210	58.5938
5	3.2215	28.9328	29.0100	20.0120	11.0449
6	2.1750	13.0345	18.0285	15.1400	26.1473
7	3.1939	3.2612	5.3869	4.0386	3.6655
8	3.5202	5.2118	5.8039	4.9386	3.9302
9	3.3299	3.6423	5.8753	5.0689	4.1797
10	3.2335	3.4663	4.9023	3.8141	3.6935
11	1.8154	29.6918	30.2174	55.5264	24.0940
12	2.8730	4.6231	4.8953	72.1241	21.2257

Critical angles obtained from deletion of specimens
Relationship between new and old principal component spaces

Table of maximum angles

Sizes of principal component-spaces being compared

Deleted sp.	1	2	3	4	5	−5	−4	−3	−2	−1
1	1.5908	3.2128	7.2555	5.1506	2.0890	3.8902	5.8600	1.5556	1.0527	0.0021
2	1.5686	2.2621	5.7630	23.4789	13.6442	18.0808	18.0881	57.7248	9.4676	0.0134
3	0.5817	3.2488	87.4599	2.8391	2.5573	11.4107	5.0989	4.9712	1.8952	0.0031
4	0.4784	1.5645	1.6599	1.6585	1.0231	3.0706	2.3138	1.8227	0.7489	0.0001
5	0.3800	0.8275	3.2435	1.1234	0.6012	1.1611	1.7366	1.7850	1.2713	0.0003
6	1.9603	4.6155	5.5704	5.4175	5.6286	23.1992	21.2229	11.8830	8.3691	0.0012
7	0.3009	3.5184	6.9671	0.8292	0.7040	1.0662	0.6104	0.6062	0.6286	0.0001
8	0.6863	3.5296	4.5290	3.6101	3.0986	4.3707	4.4058	6.5037	6.4948	0.0002
9	0.7992	1.7786	3.6186	5.6193	2.5075	3.6488	1.1529	1.042	1.0063	0.0005
10	0.4243	0.8003	4.0181	1.7125	1.4761	2.2863	1.7196	1.1254	1.0058	0.0002
11	0.4726	4.4871	5.2266	2.4627	1.5212	1.1872	0.7364	0.6416	0.4640	0.0020
12	1.5253	9.5355	9.5235	6.3116	4.7465	7.7424	8.7651	6.5108	2.7165	0.0016
13	0.8737	2.5095	6.3870	2.5784	2.4663	3.2160	2.4897	1.9326	1.8631	0.0012
14	0.7969	0.9989	3.1994	1.3997	1.3108	1.2129	1.2127	1.4007	0.4549	0.0002
15	0.1448	0.2482	0.8359	0.2391	0.2236	0.7535	0.6354	0.5532	0.5730	0.0002
16	0.2588	2.3910	2.3974	5.8733	1.6616	1.7194	2.8208	4.8985	0.9732	0.0003
17	0.4865	2.8202	25.1987	3.2744	1.0921	2.4773	1.5171	0.8185	0.0103	0.0002
18	0.7304	3.4189	3.4228	3.9027	3.3790	5.0986	5.0264	3.8483	3.7252	0.0003
19	0.2801	0.9363	3.5418	4.6230	1.0181	1.1445	1.1446	1.9684	2.1318	0.0005
20	0.0797	0.7663	1.1245	1.8452	2.7690	4.8304	5.4719	0.4495	0.4186	0.0002
21	2.4032	6.4225	11.7925	5.5467	2.2921	2.4468	3.3636	7.8417	1.7412	0.0047
22	0.1830	0.5033	2.5656	0.7626	0.6295	4.1040	2.4367	2.4476	1.9303	0.0001
23	1.0223	1.6283	3.3571	3.2367	3.3521	4.3473	2.6473	2.5399	1.4751	0.0011
24	3.5182	8.8203	22.1810	11.2810	4.6884	4.7136	3.4227	3.1152	3.1667	0.1123
25	0.2229	1.6120	4.3890	2.0263	1.6123	6.7302	4.0682	3.6053	3.6523	0.0007
26	1.2558	3.5837	21.8839	7.8848	4.3694	4.9650	3.6506	3.3997	1.6287	0.0021
27	0.1532	1.9414	12.3133	1.4477	1.4003	2.3248	4.6629	1.7438	0.8307	0.0002
28	0.1894	1.8064	1.9352	1.0988	1.0861	0.4809	0.4806	0.3745	0.3701	0.0001
29	0.2576	1.4304	1.4422	0.5978	0.5846	1.0647	1.0246	1.0382	1.3478	0.0001
30	0.2502	0.4109	0.4901	0.4844	0.4543	1.2196	1.2808	1.3800	1.0116	0.0002
31	1.2712	7.3396	9.5403	6.1540	4.1594	19.5048	15.7921	11.0968	7.4479	0.0005
32	0.3756	1.3096	1.5373	1.5311	0.6834	0.7892	1.0068	0.6714	0.0093	0.0001
33	0.2072	0.4765	5.7465	2.7584	1.1676	0.9506	1.0170	1.0593	0.8959	0.0001
34	0.2955	13.0945	17.1742	20.6944	4.0841	8.1681	4.3779	0.4830	0.0410	0.0005
35	0.1548	1.0741	4.1129	3.3843	1.7347	2.3784	2.1976	2.1938	0.3935	0.0002
36	0.4638	1.9974	4.8330	1.5610	0.9779	3.9604	1.7490	1.6149	1.7117	0.0005
37	0.2953	0.4886	2.0731	1.7096	1.2030	2.5375	2.6545	1.0328	1.0088	0.0001
38	1.0873	13.3962	69.1480	7.2624	4.8548	5.1207	7.1037	6.0903	6.4151	0.0016
39	0.0083	2.1585	88.2503	27.6896	8.3310	8.0970	5.4737	5.4481	3.2655	0.0010
40	0.4138	0.7267	5.7499	2.5725	1.8444	5.7836	5.7788	7.0171	3.8826	0.0005
41	0.9061	3.3968	60.6235	3.3278	2.3667	5.1386	3.8495	0.6575	0.1969	0.0003
42	1.4955	1.7821	3.9543	5.3184	3.7518	5.5258	4.0333	3.0189	1.7480	0.0025
43	0.2409	0.3982	1.1812	0.3016	0.2945	0.6033	0.5892	0.7086	0.6480	0.0000
44	0.2033	0.6675	0.8236	0.6480	0.6388	2.2593	2.6266	4.2227	1.1270	0.0003
45	0.6340	6.5090	14.9786	4.3535	4.2687	9.6270	8.6271	8.4441	1.4786	0.0003

46	1.8692	7.1263	21.2527	7.0744	3.9054	4.2437	3.5442	3.4068	3.5985	0.0032
47	6.8539	14.4927	16.5609	54.4452	23.2224	27.4320	46.0720	25.0455	7.0583	0.0373
48	0.8738	1.5658	17.9321	8.1964	2.7369	4.6130	3.3343	2.4730	2.1876	0.0006
49	0.0719	1.5354	3.7623	2.1407	1.0222	1.9874	1.9684	2.7471	4.7665	0.0006
50	1.9263	21.3416	55.5568	9.4406	8.1376	7.2997	6.7427	0.9579	0.9356	0.0007
51	0.7806	14.0884	28.7306	6.6604	3.5118	6.6920	9.0549	5.9537	2.8295	0.0006
52	0.5825	1.3240	3.0940	1.8149	2.1371	5.5217	5.5195	7.3392	5.9416	0.0003
53	0.4810	2.0632	4.2836	1.5774	1.4205	1.8587	1.8878	2.1553	2.5047	0.0004
54	0.5057	1.1981	1.7068	1.1105	0.9332	2.0221	3.1628	3.4793	3.9325	0.0027
55	1.1209	3.6414	14.9532	2.0780	1.6322	1.9096	2.0274	4.8757	2.5982	0.0038
56	1.6358	1.6369	12.8123	3.4369	3.0728	6.3302	6.3301	7.4174	7.0121	0.4057
57	0.8420	8.6290	9.3205	3.8377	3.8332	13.5762	12.7051	14.0230	5.7960	0.0010
58	1.8343	3.5481	44.5644	12.1577	9.8031	31.6801	47.5234	72.6773	40.2543	0.0108
59	0.9305	6.3396	42.1425	7.7512	4.5383	27.1968	15.9725	4.4247	0.2028	0.0013

Jackknifed estimates of latent roots

6.4125 1.1965 0.7027 1.1941 1.1225 0.3970 0.2199 0.2454 0.2270 0.1668 0.1155

Estimates of standard errors of latent roots (jackknifed)

1.1795 0.2215 0.1247 0.2466 0.3005 0.0684 0.0344 0.0580 0.0621 0.0292 0.0400

Jackknifed estimates of component coefficients

Component 1
−0.33741 0.31162 −0.32069 0.07121 0.25218 −0.21385 0.36196 0.35240 −0.35367 −0.35750 −0.17918 0.25448

Component 2
0.36213 −0.06663 0.24157 0.22621 0.62066 0.59795 −0.06793 0.21135 0.12496 −0.12927 −0.75148 −0.11439

Component 3
0.01450 −2.07811 −1.52413 6.82681 0.90732 −1.69271 −0.11070 0.51690 0.02723 0.15365 2.15907 −2.68128

Component 4
1.00823 −1.23640 −0.43474 3.52506 1.72502 −2.12225 0.56681 1.33273 −0.65839 −0.41043 3.30619 −4.50453

Component 5
−0.77104 −1.21413 −1.24770 −0.15508 −0.12194 2.13222 0.17946 −0.42092 −0.51854 −0.17150 0.51853 1.09372

Component 6
0.16829 −0.64668 0.34542 −0.16538 0.47896 −0.68472 −0.09574 0.00686 0.15534 −0.26250 0.07839 0.39265

Component 7
−0.89312 0.27512 0.65934 0.11492 0.13871 0.30865 −0.95880 0.49513 0.09988 −0.35192 0.51159 0.08157

Component 8
−0.64074 0.07364 −1.23321 −0.49128 0.97513 −0.07733 −0.63926 0.10573 0.65446 1.04550 −0.03426 −0.21095

Component 9
−0.09473 −0.70722 0.99403 0.14206 −0.20951 −0.12946 0.26146 0.53848 −2.21854 1.81069 −0.47172 −0.07691

Component 10
1.10051 1.03394 −0.01345 0.35004 0.72111 0.00057 −0.93013 −1.12392 −1.16744 −0.37562 0.65104 0.45047

Component 11
−1.45555 −0.10435 1.48202 0.12017 0.57964 0.00037 1.15968 −1.59988 −0.36179 0.11712 −0.26671 −0.54620

Standard errors of component coefficients (jackknifed)

Component 1
0.03655 0.05024 0.03791 0.06034 0.06947 0.05740 0.02546 0.04611 0.02413 0.02013 0.04686 0.09835

Component 2
0.10931 0.16325 0.11629 0.53625 0.15503 0.18173 0.09595 0.11013 0.09596 0.08992 0.20064 0.27603

Component 3
0.42421 0.64985 0.49364 2.13458 0.68544 0.86993 0.29637 0.53234 0.32152 0.19836 1.33288 1.81504

Component 4
0.24339 0.90966 0.64152 2.13749 0.50195 0.97111 0.22076 0.34911 0.32864 0.23599 1.06038 1.15444

Component 5
0.51127 0.55857 0.70452 0.56258 0.24155 1.21649 0.15860 0.29080 0.27092 0.19702 0.40894 0.73006

Component 6
0.32229 0.11586 0.23648 0.18404 0.13834 0.21520 0.24726 0.12683 0.18476 0.15773 0.14579 0.15344

Component 7
0.33654 0.18646 0.47900 0.23937 0.48591 0.19010 0.40222 0.14751 0.37485 0.47274 0.12075 0.15502

Component 8
0.45309 0.28935 0.52246 0.27263 0.15469 0.20951 0.41603 0.30591 0.97160 0.61165 0.29560 0.09511

Component 9
0.67770 0.40210 0.17807 0.52245 0.16141 0.58695 0.59117 0.54871 1.04015 0.45586 0.20951 0.58177

Component 10
0.70459 0.31504 0.30967 0.19202 0.37687 0.09954 0.52065 0.53279 0.95290 0.40891 0.25326 0.41362

Component 11
0.89157 0.15320 0.95364 0.09451 0.34998 0.07333 0.70907 0.98544 0.24690 0.18076 0.16623 0.35502

Estimate of number of statistically significant latent roots

No. components = 0   PRESS =   0.9831   test   stat. = 0.0000
No. components = 1   PRESS =   0.5363   test   stat. = **7.5712**
No. components = 2   PRESS =   0.5320   test   stat. = 0.0671
No. components = 3   PRESS =   0.5779   test   stat. = −0.6057
No. components = 4   PRESS =   0.5116   test   stat. = 0.8885
No. components = 5   PRESS =   0.4409   test   stat. = 0.9767

The cut-off comes here
No. components = 6   PRESS =   0.4143   test   stat. = 0.3386
No. components = 7   PRESS =   0.4115   test   stat. = 0.0306
No. components = 8   PRESS =   0.3905   test   stat. = 0.1961
No. components = 9   PRESS =   0.3744   test   stat. = 0.1190
No. components = 10   PRESS =   0.3591   test   stat. = 0.0802
No. components = 11   PRESS =   0.3420   test   stat. = 0.0478

Interpretation of the cross-validational analysis

The main things to notice in **Box 6** are:

1 There are 11 latent roots, because of the constraint. You will see that the final value in the line for latent roots is zero. For this reason, there is no twelfth principal component (although the computational routine produces one). Inspection of the array of latent vectors shows that aluminium and water are unimportant in the first principal component. In the second principal component Ti, Mn and Ca are near zero, whereas water is strongly represented. In the third principal component, ferric iron dominates completely. The first principal component is associated with most of the variation in the sample, with the second, third and fourth components being of minor importance and roughly equal in magnitude.

2 The part that lists the residuals for "variables removed" shows what happens when each variable, in turn, is deleted from the analysis in the first four principal components. This gives a good idea of whether some particular measure is really bringing essential information to the study. In the present example deleting the oxide of ferric iron (4) affects four of the principal components strongly, and in particular, components 3, 4 and 5. Removal of the oxide of Mn (6) affects the third principal component strongly, as also does H_2O (11). Removal of P (variable 12) has a pronounced effect on the fourth principal component.

3 The section headed "maximum angles" lists the residuals when each specimen of the sample is deleted in turn. The table refers to five principal components, 1 through 5. The columns headed by integers with a negative sign usually indicate atypical observations that influence the pattern of correlations (i.e. those that lie "to the side of" the main ellipsoidal body represented by the cloud of points). The columns headed by positive integers show observations, the deletion of which from the sample has an influence on the variance pattern (i.e. those that lie along the paths of principal axes, but beyond the ellipsoidal hull). The rule-of-thumb to be applied is that those observations that cause a marked increase in the magnitude of the residuals can be expected to be **atypical**, and, or, **influential**. Such observations are not always easy to detect in scatter plots of the raw data. This array of angles is quite informative. Firstly, it tells us that most of the atypicality in the sample is due to correlations, to wit, specimens 47 and 58. Specimens 2 and 6 influence both variances and covariances.

4 The section dealing with the number of principal components likely to contribute useful information indicates that five is a reasonable decision. This is also a rule-of-thumb technique, the rule being that values of the **Prediction Sums of Squares** (acronym PRESS) should at least be approximately one for a latent root to be accepted as significant. In the present example, however, the third PRESS value is very low and a case could be made for drawing the separation here.

In conclusion, we have obtained much useful information about the sample of alkaline rocks. Firstly, that all variables contribute valuable information, and that a few of them are more informative than others (the variable exclusion exercise). Secondly, we have identified several observations that exert an undue influence on the analysis (and hence the stability of the principal component elements). The investigator would be well advised to pay special attention to these divergent observations and eventually repeat the analysis without them. Thirdly, the cross-validational synthesis gives a effective indication as to the number of principal components that can be assumed to be useful. In the present case, there are five such roots. Fourthly, "jackknifed estimates" of the standard deviations for the latent roots and vectors are supplied.

Stable estimates and atypicality

The computational method of repeating a multivariate calculation many times with the exclusion of one or more observational vectors (with posterior replacement) is a useful exploratory technique for exposing lack of stability in estimates. The same data set can differ markedly from one jackknifing to the other owing to the presence of a value or values that deviate in some aspect from their fellows. This can be of quite considerable consequence if the investigator wants to reify latent vectors, or to insert them into some geometrically based procedure. For example, if the latent vectors are not stable, a latent vector with an almost zero latent root could not be expected to reflect any special algebraic relationship imposed on parts. We shall now briefly demonstrate the way in which the values can fluctuate under jackknifed sampling.

The program *jknfpca.exe* computes iteratively jackknifed estimates of latent roots and vectors of a covariance matrix or its associated correlation matrix. The trial data are in the file *Keyella.dat*. These are four standard measures on the carapace of the living ostracod species *Keijella bisanensis* from Tokyo Bay, Japan, collected in August, 1985.

Input details
Line 1: 1 the number of variables
         2 the size of the sample

Line 2: 1 type 0 for covariances, or 1 for correlations
         2 type 0 for raw data, or 1 for the logarithmic transformation

Line 3+ The data-matrix in its usual format.

Output

There are as many iterations as there are specimens (less one) in the sample. For each iteration, the jackknifed starting matrix is displayed, then its latent roots and latent vectors. A portion of the output from the *Keijella* data is shown in Table 3.

TABLE 3

Exemplification of the jackknifed estimates for the ostracod species *Keijella bisanensis*

The characters measured are length of carapace, height of carapace, distance from the antero-dorsal angle to the postero-dorsal angle, and posterior vertical height.

Jackknifed PCA for variables = 4, specimens = 20

Computations on covariances

iteration = 1

Latent roots
1 7.1885
2 3.1461
3 1.1552
4 0.0008

Latent vectors

	1	2	3	4
1	0.09122	−0.41553	−0.54955	0.71903
2	0.74530	0.48299	0.25705	0.38103
3	−0.47267	0.76587	−0.38141	0.21105
4	−0.46129	−0.08659	0.69746	0.54154

iteration = 2

jackknifed covariance matrix

	1	2	3	4
1	0.91348	−0.38700	−0.89284	−0.60116
2	−0.38700	4.69250	−1.07876	−2.39117
3	−0.89284	−1.07876	2.85354	0.87660
4	−0.60116	−2.39117	0.87660	2.15083

Latent roots
1 6.6474
2 2.8471
3 1.1157
4 0.0001

Latent vectors

	1	2	3	4
1	0.05353	−0.45449	−0.52393	0.71838
2	0.79137	0.39940	0.25968	0.38311
3	−0.35238	0.79586	−0.44884	0.20242
4	−0.49668	0.02275	0.67572	0.54423

iteration = 3

jackknifed Covariance matrix

	1	2	3	4
1	0.95307	−0.31149	−1.07360	−0.62532
2	−0.31149	4.66008	−1.58082	−2.25165
3	−1.07360	−1.58082	3.55231	1.15041
4	−0.62532	−2.25165	1.15041	1.97132

Latent roots
1   7.0820
2   2.9821
3   1.0710
4   0.0016

Latent vectors

	1	2	3	4
1	0.09665	−0.44558	−0.52393	0.71945
2	0.73254	0.52109	0.21555	0.38130
3	−0.50344	0.72553	−0.41847	0.21223
4	−0.44788	−0.05940	0.70987	0.54034

etc.

PRINCIPAL COMPONENT ANALYSIS OF COMPOSITIONAL DATA

The calculations proceed in the same manner as outlined in the section on principal component analysis, but using now the centred log-ratio covariance matrix or its correlational counterpart, introduced earlier on. One may also use the log-ratio covariance matrix, but for the present purpose, the former presents interpretational advantages and is indeed the version promoted by Aitchison (1986). In many published geochemical applications, the "crude" covariances or correlations are used, but this approach suffers from the crippling defect associated with interpreting crude covariance structures in a statistically appropriate way.

Log-contrast of a D-part composition

The log-linear contrast of the components of the composition $\mathbf{x}$ is any combination

$$a_1 \log x_1 + \ldots + a_D \log x_D \qquad (3:14)$$

$$= \mathbf{a}^T \log \mathbf{x}.$$

Note, that

$$a_1 + \ldots + a_D = \mathbf{a}^T \mathbf{j} = 0.$$

This latter property ensures that a log contrast can always be expressed as a linear combination of log-ratios with a common component divisor (**j** denotes the unit vector). For example,

$$\mathbf{a}^T \log \mathbf{x} = \mathbf{a}^T \log\{\mathbf{x}/g(\mathbf{x})\}. \qquad (3:15)$$

Two log-contrasts, $\mathbf{a}^T \log \mathbf{x}$ and $\mathbf{b}^T \log \mathbf{x}$, are orthogonal if $\mathbf{a}^T \mathbf{b} = 0$.

The appropriate formulation of the principal component solution for compositional data is then to find the latent roots and vectors satisfying

$$(\Gamma - \lambda_i \mathbf{I})\mathbf{a}_i = \mathbf{0} \qquad (3:16)$$

where Γ is the centred log-ratio covariance matrix.

The log-contrast vector $\mathbf{a}_i^T \log \mathbf{x}$ is called the i-th log-contrast principal component. For the purposes of constrained principal components, Aitchison (1986) recommends the use of the centred log-ratio covariances and this is what we use here. We shall now look at a simple example, the Haitian bolide data again.

Try now the program *pcaconst.exe* with data on bentonites stored in the file *haitipc.dat*. You do this by typing at the *DOS*-prompt:

C:\pcaconst‹haitipc.dat

This program performs a constrained principal component analysis on a specified number of parts. It also provides a principal component analysis of the raw data in order to indicate what happens when the inappropriate method is used. The trial data in *haitipc.dat* are chemical elements determined on impact glass samples from the Late Cretaceous bolide impact on Haiti (Sigurdsson et al., 1991) and already introduced. The data are determinations on the oxides of Si, Al, Mg, Ca, Na, K and S.

Instructions for using the program **pcaconst**

Line 1: Specify the dimensionality of the problem (i.e. the number of parts)

Line 2: The title of the job

Line 3: sample size

Line 4 and following: the data-matrix.

Output details

Save the Haitian example, and get it up on the monitor screen. The output of immediate interest begins with the centred log-ratio data-matrix, followed by the centred log-ratio covariances. The latent roots and vectors for the constrained and raw principal component analyses and the scatter diagrams of the scores for the former are computed by the program. There are hardly any differences between the first principal components of the raw and constrained analyses. There are some differences in the second latent vectors, but it is not until the third latent vector that substantial conflicts begin to appear.

There are six non-zero latent roots (remember that the constraint made one of them zero since the rank of the centred log-ratio covariance matrix is one less than the number of parts). Hence, the seventh latent vector is to be excluded. In the case of the raw data-matrix, the smallest latent root is almost zero. The small value obtained is due to rounding errors.

There is a second data-matrix for you to try, to wit, *atlant.dat*, a set of chemical analyses of alkaline rocks, also used in an earlier example in **Box 6** (data from Borley, 1974; Sørensen, 1974). As an exercise in reification, compare the two sets of answers you will be given by the program, one for the appropriate model, and one for the inappropriate one.

Q-MODE ANALYSIS

The next class of methods to be illustrated is that of *Q*-mode analysis (Q comes before R in the alphabet, hence the seemingly cryptic designation the usage derives from the early psychometricians). A *Q*-mode analysis is concerned with probing relationships between the objects of a sample. The reason for wanting to do this has not always been well understood, nor appreciated by statisticians, who have, in the past, tended to regard it as a furtive procedure, best left unsung. This attitude has begun to change and there are few statisticians today who would negate the usefulness of "inverted factor analysis" for some aspects of data-analysis. The procedure we recommend here is that of *Principal Coordinate Analysis*, developed by Gower (1966) in an attempt to provide a mathematically sound treatment of the associations between columns in a data matrix. We usually rely on this method because John Gower specifically designed it to accommodate the three classes of variables: quantitative, dichotomous, and qualitative. In this particular respect, it is superior to its competitors, at least in geological and biological work, since rival techniques do not possess this quality. Reyment and Jöreskog (1993) provide an account of the area of *Q*-mode analysis (Chapter 5 in that book) to which you are referred for a more complete coverage than we aim at here, including what is rather inaccurately known as *Q*-mode factor analysis, and its appurtenances, but which is a variant that enjoys considerable popularity in geochemical circles.

It may be necessary to stress that it is not a legitimate procedure to attempt to interpret, that is, reify, the "loadings" of a Q-mode analysis, although we are well aware that this verges on being common practice in geochemistry and petrology. Examples have been occurring in recent biological literature. The otherwise well conceived paper by Thiede et al. (1997) is based on inappropriate reifications of Q-mode analyses. They used a computer program often invoked in bio-oceanology, but the scientific basis for which is by no means unchallengeable We shall briefly examine the issues involved. The article deals with oceanic surface conditions on the sea floor of the southwest Pacific Ocean. Thiede et al. (1997) analysed more than 180 surface samples of sediments from the southwest Pacific Ocean with respect to their contents of foraminifers, opaline material and sedimentary types. The aim of the study was to attempt to unravel oceanic conditions at the surface of the sea above the sampling sites.

The 'statistical' evaluation of the data was made using a method known as Q-mode factor analysis with varimax rotation. This is a fixed-mode method of data-analysis (cf. Reyment and Jöreskog, 1993 for a description of the method). It is not factor analysis in the true sense and can be most conveniently described as inverted principal component analysis (that is the extraction of the latent roots and vectors of a matrix of associations in the space of the objects) followed by a rotation of the axes. As a purely graphically device, Q-mode factor analysis can be expected to give a reasonable representation of the relationships between the objects of a data-set. As is, however, often the case in geology, Thiede et al. (1997) went on to reify the elements of the 'factor matrix', which is neither advisable nor statistically permissible as has already been pointed out. Among the many objections that can be raised, the following may be mentioned:

The high probability of instability of the latent vector elements under repeated sampling. This source of potential inaccuracy could probably have been somewhat minimized if the estimates had been based on some method of repeated sampling, such as jackknifing the input and repeating the computations. Cross-validation is a suitable procedure for achieving this goal. Whichever amelioration is used, there is no altering the fact that Q-mode factor analysis is not a statistical method but rather a fixed-mode data-analytical proxy applicable solely to the sampled material for which it was computed. All is not lost, however. The method used by Thiede et al. (1997) could easily be updated into a statistically valid procedure.

Gower's Assocation Measure

Gower's measure of similarity between objects i and j is defined as the absolute difference between them for variable p, divided by the range of the variable and the sum for each variable subtracted from one (Digby and Kempton, 1987, p. 20). In a situation where all variables are of the same kind and the association is of the correlation matrix kind, the Eckart-Young theorem, to be introduced below on p. 141, applies and there may be no real advantage that makes one mode more effective than the other in constructing an ordination. In cases where the variables are mixed, and product-moment correlations are not applicable, the singular value decomposition cannot be applied and the usual method of extracting latent roots and vectors must be employed.

For our present purposes, the essential features of the method can be summarized in the following words:

The usual situation is that given a set of N points in p-dimensional space (i.e. the R-space spanned by the variables), with coordinates given by the rows of the data-matrix for each of the points, the squared Euclidean distance d_{ij} between any pair of points, P_i and P_j, is expressed by eqn. (3:17):

$$d_{ij} = \sum_{k=1}^{p}(x_{ik} - x_{jk})^2 \qquad (3:17)$$

The situation addressed by principal coordinate analysis is the converse of that described by eqn. (3:2) in that, if you have a set of N values that represent the squared distances within a set of N points in some Euclidean space (the Q-space spanned by the observations), how do you go about finding the coordinates of these points? The idea is by no means new, having been around since 1935 (Schoenberg, 1935) under various designations (Gordon, 1981). It was, however, Gower (1966) who made the method more generally known. The steps for doing the calculations are:

1. Make the centroid of the points lie at the origin of the coordinates.
2. Transform the matrix **D** of squared distances into a matrix **A** of inner products

$$a_{ij} = -\tfrac{1}{2}[d_{ij} - d_{i.} - d_{.j} + d_{..}]. \qquad (3:18)$$

3. Obtain the values of the coordinates x_{ij} from the distances d_{ij} by a standard principal component extraction – i.e. a standard extraction of latent roots and vectors. This can be done, since **A** is a real symmetric matrix and can therefore be diagonalized

$$\mathbf{A} = \mathbf{V}\Lambda\mathbf{V}^\mathsf{T} \qquad (3:19)$$

4. The columns of the orthogonal matrix **V** contain the latent vectors ($\mathbf{v}_1$, $\mathbf{v}_2$, ..., $\mathbf{v}_N$) corresponding to the latent roots (λ_1, λ_2, ..., λ_N).
5. The matrix of coordinates **X** is found by computing

$$\mathbf{X} = \mathbf{V}\Lambda^{1/2} \qquad (3:20)$$

6 Usually only the first 2 to 3 coordinates attract much interest (that is, $k = 2$ or 3). The sum of the remaining latent roots, expressed as a percentage of the trace (the sum of all of the latent roots) supplies an indicator of how well the analysis has succeeded in preserving distances between points. If the residual is very large, then caution is advisable on interpreting the ordination obtained, as there is much random variation in the data. The usual method of computing principal coordinates is to be found in the program entitled *pcoord.exe*. There are a great number of alternatives in applying this program, owing to the comprehensiveness of the field it attempts to cover.

Instructions for using the program **pcoord**

Line 1: The title of the job

Line 2: information to be supplied
    1 size of the sample (maximum of 240 specimens)
    2 number of variables
    3 number of latent roots (maximum of 9)
    4 number of "quantitatives" = continuous variables
    5 number of qualitative variables
    6 number of dichotomous variables (presence–absence data)

These entries are only valid for Gower's metric below; if it is not selected, enter noughts here.

    7 0 gives Gower's metric
        1 gives the usual correlation coefficient
        2 gives Lance's metric
        3 gives the Euclidean metric
    8 1 for a listing of the data-matrix, otherwise 0.
    9 0 if the data matrix is in the usual form of N, p. If it is available as an p, N matrix, type 1
   10 1 for the minimum spanning tree. Please note that the minimum spanning tree is only really useful for relatively few observations, say around 30 at the most. If you have a large sample, the superimposition of the tree will not be enlightening owing to cluttering of the graph.

Line 3: If dichotomous data occur the default entry for such data is +1 for a character that is present and 1 if it is absent.

Lines 4 and following contain the data-matrix

There are several similarity and dissimilarity coefficients available. One of these by Lance is mentioned but not exemplified; any text on classification will tell you more about it should you be interested (Lance and Williams, 1965; Anderberg, 1973).

The example presented in **Box 7** is run by invoking the program *pcoord* to be used in conjunction with the data-set *afrocrd.dat*, the set of seven 'quantitatives', the dimensions of the test of the Nigerian Campano–Maastrichtian foraminiferal species, *Afrobolivina afra* REYMENT, also used for exemplifying principal component analysis. These variables are all of the quantitative class (as opposed to qualitative and dichotomous).

Box 7: Principal coordinate calculations for *Afrobolivina afra*

Program: **pcoord**

Data: **afrocrd.dat**

The *Afrobolivina afra* data

Variables = 7 Individuals = 70

Gower's Principal Coordinate Analysis

Latent roots (first nine) of transformed association matrix

5.6550
1.9199
1.4091
1.1559
0.8573
0.7597
0.5595
0.5025
0.4603

These are the first nine latent roots of the 70 × 70 association matrix. Note that the first root is very much larger than the others, but the fall-off thereafter is rather slow. There are "only" nine listed latent roots, because this is the greatest number saved by the program. There is absolutely no point in computing and saving all the roots in a Q-mode procedure. Generally speaking, if the fall-off is slight for all roots, then it is likely that there is much random variation in the data. This indicates that a Q-mode analysis is not going to be useful.

Specimen Coordinates

1	−0.28348	0.01638	−0.05021	0.25522	−0.15939	0.00604	−0.02993	−0.04272	−0.11962
2	0.25865	0.00721	−0.23345	0.05159	−0.23134	0.20498	0.11087	0.07184	−0.03053
3	0.09120	0.04029	−0.09914	0.05119	−0.11538	−0.13960	−0.10735	0.10258	−0.03324
4	0.28823	−0.10363	−0.02184	−0.09206	−0.03907	−0.01850	−0.06515	−0.10187	−0.12495
...	...	...	...	...					
69	−0.15529	0.25355	0.09627	−0.22640	0.05898	0.01644	−0.14754	−0.04719	0.08388
70	0.01783	0.26671	0.19010	−0.01661	0.08717	0.08301	−0.18257	−0.01753	0.07600

These values are the principal coordinates to be plotted by means of *Graph Server*. They are stored in the file **prcrd.**

The primary and perhaps sole interest in doing a principal coordinate analysis lies with the graphical representation of relationships between objects. (N.B. in some geochemical applications of inverted principal component analysis – known as Q-mode factor analysis, attempts are made at reifying (interpreting) the elements of the latent vectors; please, never be tempted to anything so utterly meaningless, even in the face of peer-pressure.)

An example of the plot produced by *Graph Server* for principal coordinates is displayed in Fig. 28. These data are the same as were used for exemplifying the output from the program for principal components (Fig. 27(a) and (b)). Note that the plot brings out more evidence of structure in the data-set than did the principal component results. Fig. 28 brings to the fore the main merit of Q-mode analysis, notably, the possibility offered of uncovering meaningful grouping in a data-set. The result illustrated in Fig. 28 could be used for subdividing the data into categories for subsequent detailed analysis.

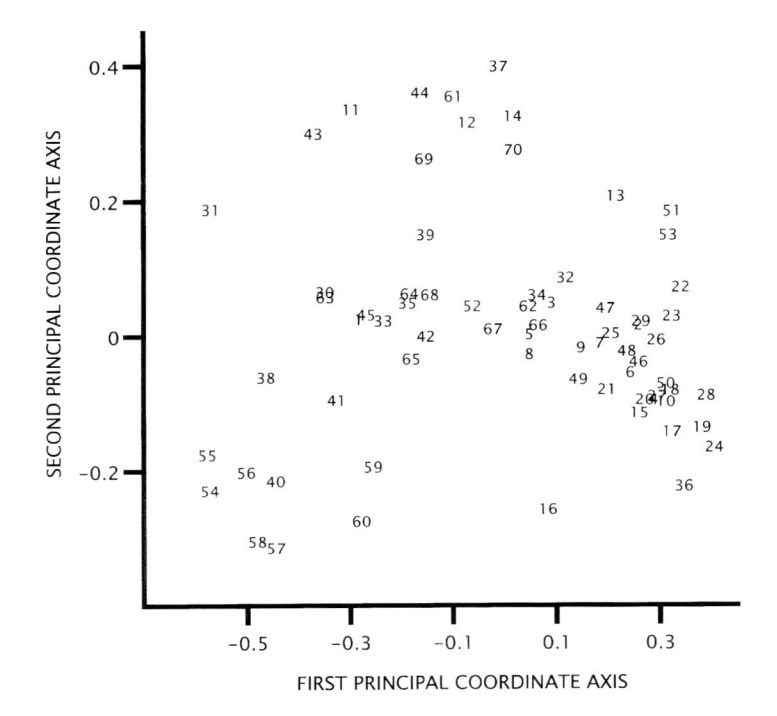

Fig. 28. Plot of the first two principal coordinates for the *Afrobolivina afra* foraminiferal data. The points do not form a tight cluster (70 specimens) which is due to polymorphism in the size of the proloculus and attendant morphological differences. There are at least three natural groups represented.

PRINCIPAL COORDINATES FOR COMPOSITIONAL DATA

A log-contrast principal coordinate analysis can be obtained as the dual of logcontrast principal components for compositional data. The concept of distance between two compositions $\mathbf{x}_1$ and $\mathbf{x}_2$ is defined as

$$\sqrt{\sum_{i=1}^{D}([\log(x_{1i}/g(\mathbf{x}_1)) - \log(x_{2i}/g(\mathbf{x}_2)])^2}$$

This rendition of principal coordinates only takes into account statistical "distances", i.e. the "quantitatives" of Gower (1966, 1971). Gower's "qualitatives" and "dichotomies" would seem to be available without difficulty should this be found necessary in a particular study such as could occur in environmetrics.

Aitchison (1986) advocates the centred log-ratio covariance matrix for use with principal coordinates. We provide here a modification of this suggestion by means of the centred log-ratio matrix of associations of Gower (1971), judging this to be more in sympathy with the original intentions of the author of the method. This is available in the program *pcrdcons.exe*, an adaptation of program *pcoord.exe*, and applies only to a data matrix of parts. The instructions for using this program are:

Instructions for using the program **pcrdcons**

Line 1: The title of the job.

Line 2: 1 the number of rows in the compositional data-matrix;
        2 the number of columns in the matrix (the number of chemical parts);
        3 the number of latent roots (at least 3, at most 9);
        4 1 for symbols if required for line-plots (and full output)
        5 1 for full output, otherwise 0

Line 3 and following, the data-matrix arranged as rows by columns (the usual format of a table of chemical compositions).

Output

The full output from the program provides the matrix of associations, latent roots and vectors (the coordinates) and pairwise plots of the coordinates. The residuals after fitting two, then three, coordinate representations of the distances between points are listed at the end of the output. Large residuals are indicative of a poor fit. You can try an example. Type:

pcrdcons‹canary.dat

which is an analysis of chemical determinations on alkaline rocks from the Canary Islands, used already several times (Sørensen, 1974). The coordinates are saved in a file called **aitprcrd** for insertion into *Graph Server*.

MORE ON R- AND Q-MODE METHODS

A fundamental theorem of multivariate analysis is the *singular value decomposition*, first enunciated in 1889 by the celebrated English mathematician J. J. Sylvester, expanded in scope by L. Autonne in 1913, and generalized to rectangular matrices by Eckart and Young (1936) (Eckart was a physicist, Young a psychometrician). It is the key to the analysis of *QR*-mode relationships in the multivariate population. In statistics, the term often used (at least in psychometrics) is the *basic structure of a rectangular matrix* (cf. Reyment and Jöreskog, 1993).

Let $\mathbf{X}$ be a given data-matrix of order N by p, where $N > p$, and let r be the rank of $\mathbf{X}$. In most cases, $r = p$, but we have already learned that the log-ratio data matrix is of rank $r = p - 1$. The singular value decomposition of $\mathbf{X}$ states that

$$\mathbf{X} = \mathbf{V}\mathbf{\Gamma}\mathbf{U}^T \qquad (3:21)$$

where $\mathbf{V}_{(N,r)}$ is a matrix with orthonormal columns, $\mathbf{U}_{(p,r)}$ is orthonormal, and $\mathbf{\Gamma}_{(r,r)}$ is a diagonal matrix with r positive diagonal elements $\gamma_1 > \gamma_2 > \ldots > \gamma_r$. These gammas are called the *singular values* of $\mathbf{X}$.

Interpretation of the singular value decomposition

1 The product $\mathbf{X}\mathbf{X}^T$ of a data-matrix by itself is sometimes referred to as the *major product moment*. It is of order N by N and has r positive latent roots $\gamma_1^2, \ldots, \gamma_r^2$ and $(N - r)$ zero latent roots. The corresponding latent vectors are $\mathbf{v}_1, \ldots, \mathbf{v}_r$.

2 The alternative multiplication of a data matrix by itself, $\mathbf{X}^T\mathbf{X}$, called the *minor product moment*, in the terminology extant in factor analysis, is of order p by p. It has r positive latent roots, $\gamma_1^2, \ldots, \gamma_r^2$. These are the same as for the major product moment. There are $(p - r)$ zero latent roots. The latent vectors corresponding to the positive latent roots are $\mathbf{u}_1, \ldots, \mathbf{u}_r$.

3 If $\mathbf{v}_m$ is a latent vector of $\mathbf{X}\mathbf{X}^T$ and $\mathbf{u}_m$ is a latent vector of $\mathbf{X}^T\mathbf{X}$, both corresponding to the latent root γ_m^2, then the following relationships hold:

$$\mathbf{v}_m = (1/\gamma_m)\mathbf{X}\mathbf{u}_m \qquad (3:22)$$

and

$$\mathbf{u}_m = (1/\gamma_m)\mathbf{X}^T\mathbf{v}_m. \qquad (3:23)$$

Thus, there is an easy path from the *Q*-mode state to the *R*-mode one.

4 More generally,

$$V = XU\Gamma^{-1}$$

and

$$U = X^T V \Gamma^{-1}. \qquad (3:24)$$

The singular value decomposition of a rectangular matrix is the best method for programming *Correspondence Analysis* and the *biplot* (see Chapter 4). The biplot is rather like the idea embodied in correspondence analysis but is more general in scope (Gabriel, 1995a,b). Gabriel (1971) proposed it originally for continuously distributed variables but the method has since then been vastly expanded so as to encompass data in the form of contingency tables (Gabriel, 1995a,b). It was originally exemplified by climatological data in Gabriel's paper, which clearly betokens its geoscientific importance.

As indicated by condition (3:24) above, you can work on very large data-matrices, on which relatively few variables have been determined, by the following algorithm:

1 Compute the minor product moment, $X^T X$. This is p by p and $p < N$.
2 Compute the positive latent roots of the minor product moment and the corresponding latent vectors.
3 Compute then the vectors V from the first part of condition (3:24).

These vectors are of interest in Q-mode analysis in that they constitute the coordinates sought for ordinating sample points.

It can be very instructive to see how the singular value decomposition actually works on a data-matrix. We have a little program that demonstrates this succinctly to which we shall now introduce you.

Type *singval* at the *DOS*-prompt, and follow the instructions that appear on the monitor. This program operates on the data sets in *matrx3.dat*.

We shall first inspect some of the output. You can do this yourself just by typing:

C:\singval>output.sav

and then going through the file *output.sav*, using an editing facility. Look now at the contents of **Box 8**.

Box 8: Exemplification of the singular value decomposition

Program: **Singval**

Decomposition of data-matrices:

Original matrix:

$$\begin{bmatrix} 117.3 & 99.8 & 97.3 \\ 93.3 & 81.8 & 81.1 \\ 76.6 & 71.3 & 71.8 \\ 62.3 & 59.7 & 59.9 \\ 42.1 & 42.5 & 42.5 \end{bmatrix}$$

(matrix taken from Reyment and Jöreskog, 1993, p. 5; please note that the following listings reproduce the computer output with respect to number of decimals and no such literal accuracy is implied)

Matrix **U**

$$\begin{bmatrix} -0.6152 & 0.6035 & 0.3958 \\ -0.5008 & 0.1517 & -0.4129 \\ -0.4286 & -0.3649 & -0.5275 \\ -0.3546 & -0.4575 & 0.1013 \\ -0.2474 & -0.5199 & 0.6199 \end{bmatrix}$$

Diagonal matrix

295.9917   8.5579   0.6758

Matrix $\mathbf{V}^T$

$$\begin{bmatrix} -0.6224 & -0.5561 & -0.5507 \\ 0.7727 & -0.3247 & -0.5454 \\ -0.1245 & 0.7650 & -0.6318 \end{bmatrix}$$

Check product against original matrix:

Product $\mathbf{UDV}^T$

$$\begin{bmatrix} 117.3 & 99.8 & 97.3 \\ 93.3 & 81.8 & 81.1 \\ 76.6 & 71.3 & 71.8 \\ 62.3 & 59.7 & 59.9 \\ 42.1 & 42.5 & 42.5 \end{bmatrix}$$

What we have now done is decompose the input matrix with five rows and three columns into

1 a diagonal matrix,
2 a 5 × 3 matrix **U** of orthogonal vectors, and
3 a 3 × 3 matrix of orthogonal vectors, **V**.

The former is the *Q*-mode part of the decomposition (that is, relevant to the sample size) and the latter is the *R*-mode part, relevant to the number of variables. It is easy to see that **V** contains the **principal components** of the data matrix, and **U** the "**principal coordinates**". The next step provides a check on whether the computations are correct or not. The product of the three matrices should bring back the original data-matrix.

Check product against the input matrix:

Product **UDV**^{T}:

$$\begin{bmatrix} 117.3 & 99.8 & 97.3 \\ 93.3 & 81.8 & 81.1 \\ 76.6 & 71.3 & 71.8 \\ 62.3 & 59.7 & 59.9 \\ 42.1 & 42.5 & 42.5 \end{bmatrix}$$

A second extract from the same source as the foregoing example should suffice to bring home the great value of the singular value decomposition in applied multivariate statistics. You will also appreciate, no doubt, that by an appropriate method of scaling axes, such as is achieved in Correspondence Analysis, the properties of a data-matrix can be elegantly exposed in just a few diagrams.

Matrix **U**

$$\begin{bmatrix} -0.4913 & 0.7135 & 0.3864 & 0.0129 & -0.3164 \\ -0.4625 & 0.2379 & -0.4173 & -0.0001 & 0.7451 \\ -0.4883 & -0.2836 & -0.5249 & 0.3861 & -0.5065 \\ -0.4348 & -0.3814 & 0.1052 & -0.8040 & -0.0892 \\ -0.3423 & -0.4566 & 0.6244 & 0.4520 & 0.2830 \end{bmatrix}$$

Diagonal matrix

270.9985   63.1665   7.8758   0.0000   0.0294

Matrix $\mathbf{V}^\mathsf{T}$

$$\begin{bmatrix} -0.6779 & -0.3325 & -0.3776 & -0.3499 & -0.4061 \\ 0.6942 & -0.4993 & -0.1561 & -0.4395 & -0.2263 \\ -0.1779 & -0.2537 & 0.9084 & -0.2525 & -0.1223 \\ 0.1612 & 0.5641 & 0.0806 & -0.0000 & -0.8058 \\ 0.0292 & -0.5075 & 0.0365 & 0.7878 & -0.3458 \end{bmatrix}$$

The complete set of test-data is housed in the file *matrx3.dat*. If you want to use the program *singval.exe* for experimenting on matrices of your own, just insert your material into a file you call *matrx3.dat*, using exactly the format you encounter in that file and invoke the program as before. Before doing this, however, we suggest you copy the original file *matrx3.dat* into another subregister.

ANALYSIS OF CORRESPONDENCES

The rather cryptic designation is the inaccurate English rendition of "Analyse des Correspondances". A more exact translation would have been "Analysis of Associations". Since the Analysis of Correspondences was developed as a graphical technique (Benzécri, 1973) much has happened. For example, Greenacre (1984) gave a more general treatment of the subject, including practical applications. Recently, Gabriel (1995a,b) has placed Simple Correspondence Analysis and Multiple Correspondence Analysis (Hill, 1974) in the same context as the Biplot and shown that they must now be regarded as less efficient procedures for analysing contingency tables. Gabriel (1995a, p. 210), advocating discontinuation of the use of the technique, concluded that the plots obtained from multiple correspondence analysis fail to portray important features of the data and that they do not display the association between objects and categories reliably. In a second article (Gabriel, 1995b), it was pointed out that the superimposition of two sets on the same plot in simple correspondence analysis is not warranted because there is no row-to-distance interpretation that can be justified and that this practice should therefore be discontinued. The joint display of points on the same graph has no virtues over and above those of the separate row-based and column-based plots, according to Gabriel's researches. Gabriel's several arguments appear convincing and it remains to see how the future fate of correspondence analysis will unfold. In the present connection, we confine interest to an example of simple correspondence analysis, *sensu* Benzécri (1973). An important and very comprehensive reference for the broader field of correspondence analysis is the book by Greenacre (1984).

The program **benzec.exe** can now be invoked for doing a correspondence analysis, using the original formulation proposed by J.-P. Benzécri (1973). The trial data are in the file *fenantrn.dat*, which are 36 samples of oils from the North Sea (Norway), upon which 13 phenanthrene compounds have been determined (Telnaes et al., 1987).

Instructions for using the program **benzec**

Line 1: The title of the job.

Line 2: 1 number of variables ($<=50$)
        2 number of observations ($<=440$)
        3 put 0 for raw data
          1 for $\log_e$ transformation of the data
        4 0 lists the input data
          1 suppresses listing of the data

Line 3+: The data-matrix then follows in free format.

Last line: For requested data listing, supply symbols for the variables (one letter or number per variable).

Output details

1 The similarity matrix
2 Sums of rows and columns
3 Latent roots and vectors
4 Coordinates of the correspondence analysis for variables and specimens. This output is saved in the file *bencrd* for insertion into *Graph Server*

A succession of bivariate plots showing the variables and specimens located on the same plot is the principle aim of the analysis. You can see some of the output by consulting **Box 9**. Type the following at the *DOS*-prompt.

c:**benzec**<fenantrn.dat

    Box 9: Example of correspondence analysis

    Program: **benzec**

    Data: **fenantrn.dat**

Analysis of correspondences 147

The phenanthrene data from Telnaes et al. (1987) analysed by correspondence analysis. The data consist of 11 phenanthrene determinations made on 36 samples of North Sea oils, to wit, monomethyl phenanthrenes (4 species), dimethyl phenanthrenes (6 species) and phenanthrene.

The sums of the rows are all about 100; this is quite in order for the requirements of correspondence analysis, since the method was specifically devised for the analysis of contingency tables, the entries in which may be interpreted as probabilities, and hence have a constant sum.

One latent root is lost because of the effects of the contingency table and the effect of scaling (there is an analogy in principal coordinate analysis).

No.	Latent root	percent	cumulative percent
2	0.42829E-02	47.877	47.877
3	0.18375E-02	20.540	68.417
4	0.13171E-02	14.723	83.140
5	0.46818E-03	5.234	88.373
6	0.41020E-03	4.586	92.959
7	0.30684E-03	3.430	96.389
8	0.14440E-03	1.614	98.003
9	0.10267E-03	1.148	99.151

variable projections

1	−0.000	−0.071	0.004	−0.019	−0.006	0.020	−0.006	0.002
2	0.080	−0.022	−0.036	0.023	−0.034	−0.023	0.006	0.008
3	0.047	−0.013	0.026	0.002	0.014	−0.019	0.000	−0.002
4	−0.055	−0.002	−0.066	0.008	0.021	0.004	0.009	−0.004
5	−0.085	−0.027	0.042	0.022	0.003	−0.017	−0.009	−0.002
6	0.201	−0.015	0.059	−0.004	0.025	0.014	0.037	−0.019
7	0.136	0.055	−0.007	0.048	0.033	0.031	−0.033	0.008
8	0.004	0.053	−0.010	−0.020	−0.022	0.003	−0.006	−0.005
9	−0.009	0.043	0.018	−0.033	0.026	−0.015	0.005	0.023
10	−0.027	0.033	0.023	−0.004	0.003	−0.010	−0.004	−0.022
11	−0.092	0.050	0.059	0.038	−0.025	0.038	0.024	0.011

sample projections

1	−0.092	−0.017	0.018	−0.004	−0.003	−0.030	−0.001	−0.007
2	−0.085	−0.017	0.029	−0.008	−0.008	−0.024	−0.006	−0.001
3	0.015	−0.030	−0.010	0.019	0.027	−0.002	−0.009	0.007
4	0.007	−0.034	0.023	−0.007	−0.021	−0.017	−0.007	−0.012

....

34	−0.110	0.043	−0.000	0.019	−0.001	−0.032	−0.010	0.009
35	−0.087	−0.020	−0.006	0.030	−0.007	0.020	0.026	0.014
36	−0.078	−0.002	0.005	0.015	0.007	0.004	−0.020	0.006

This listing is just the values for variables and specimens combined. It is in this form the results are saved for plotting in the file *bencrd*. Please note that the use of the word 'variables' in the present connection is not quite correct for all situations that can occur in that in cases where we are not dealing with absolute measurements, the correct designation is 'parts', but you already know this.

Interpretation of the correspondence analysis of the phenanthrenes

The mode of interpretation of correspondence analysis is entirely graphical in that the reason for doing the analysis is to

(a) Look for clustering in the data;
(b) Look for a relationship between the variables (parts) and the data-points.

It is this pairing of aims – variables coupled to observations – that justifies the designation – QR-method.

When you run the test-data, you should think about examining the printout of the simple scatter diagrams for such relationships. That is, you should look for groupings of specimens that occur in proximity to one or more variables – this is interpretable as indicating to what extent parts have an effect on certain specimens. The scaling procedure was constructed to place specimens influenced by particular variables in the immediate vicinity of them.

THE BIPLOT

Although the biplot of Gabriel (1971) may seem to be just a simple variant of correspondence analysis due to the use made in both cases of the singular value decomposition, it is, in effect, not. As already mentioned, Gabriel (1995a,b) has shown the biplot to be a more general way of treating the rows and columns of a contingency table. Biplots provide graphs of the N observations as well as the relative positions of the p variables, superimposed in the same figure. As originally conceived (hence the name) the technique was meant to be applied to just a single two-dimensional plot, under the assumption that the data-matrix had rank 2 (or almost rank 2). However, the extension to further dimensions was quickly made by the innovator of the method. The usefulness of the original concept of the biplot is that the results are easy to interpret and it can be easily incorporated into a wider field of multivariate analysis (Gordon, 1981).

The basic concept in the formulation of the biplot is the fact that the decomposition of the (rectangular) data-matrix can be done in two ways. Firstly, there is the application of the singular value decomposition in the manner already demonstrated in **Box 8**.

$$\mathbf{X} = \mathbf{V}\mathbf{\Gamma}\mathbf{U}^T.$$

The second way is by the breakdown into a non-unique product of two matrices

$$\mathbf{X} = \mathbf{G}\mathbf{H}^T \tag{3:25}$$

where $\mathbf{G}$ is $N \times r$ and $\mathbf{H}$ is $p \times r$. Here, N is the sample size (the number of rows in the data-matrix, p is the number of variables, and r is the rank of the data-matrix. In computing the biplot, it is convenient to "standardize" $\mathbf{X}$ so that the mean on each variable is 0.

If the rank of $\mathbf{X}$ is exactly 2, the vectors constituting $\mathbf{G}$ and $\mathbf{H}$, $\mathbf{g}_i$ and $\mathbf{h}_j$, respectively, yield the biplot. If $r > 2$, the singular value decomposition can be invoked to construct a matrix $\mathbf{X}_2$ of rank 2 which is the best approximation to $\mathbf{X}$ in the sense that the sum of squares of the elements of $(\mathbf{X} - \mathbf{X}_2)$ is a minimum. The indeterminacy in $\mathbf{G}$ and $\mathbf{H}$ can be removed by replacing each of the vectors forming $\mathbf{G}$ by a point located at the end of them. A further manipulation can be made which ensures that the distance between the i-th and j-th points in the geometrical representation is equal to the "distance" between the objects in the i-th and j-th rows of $\mathbf{X}$. This is done by defining $\mathbf{G}$ as

$$\mathbf{G}\mathbf{G}^T = \mathbf{X}_2 \mathbf{M} \mathbf{X}_2^T \tag{3:26}$$

where the matrix $\mathbf{M}$ specifies the metric used on the rows of $\mathbf{X}_2$.

If $\mathbf{M} = \mathbf{I}_2$, the identity matrix, the distance between the points at the end of each pair of **g**-vectors corresponds to the Euclidean distance.

If $\mathbf{M} = \mathbf{S}_2^{-1}$, the inverse of the sample covariance matrix for the reduced data-matrix $\mathbf{X}_2$, then the Euclidean distance between the points of each pair of **g**-vectors corresponds to the Mahalanobis generalized statistical distance between the corresponding objects in $\mathbf{X}_2$. (The Mahalanobis distance is defined in the next chapter.)

These two special properties of Gabriel's biplot make the technique attractive in theoretical advancements, for example, in the spline-based image-analytical studies of shape-variation, and other developments in multivariate statistics.

The standard deviation of the k-th variable is given by the length of $\mathbf{h}_k$ and the correlation between any two variables is expressed by the cosine of the angle between them. More detailed discussions of the biplot and its many applications are to be found in Gordon (1981), Jackson (1991), Jolliffe (1986) and Gabriel (1995a,b).

Instructions for using the program **Gabriel**

Line 1: 1 sample size
        2 number of variables
        3 if full output is required, type 0
           if only the graphical output for later plotting is needed, type 1

Line 2: The title of the job

Line 3 and following, the data-matrix in its usual free format

The file *biplots* contains the output for graphing by means of *Graph Server*. If the full-output option has been chosen, the program also gives a screen-plot as a guide.

The simple example in **Box 10** illustrates the output from program *gabriel.exe* using data in the file *gabriel.dat*. An interpretation of the results is given in the box.

There is a more comprehensive example for you to try at leisure. The file is for data on the environmetric analysis of the *Leptocythere sp.* ostracod data in the file designated as *leptobip.dat*, this being occurrences of that ostracod observed at three stations (see **Box 13** for further details regarding the nature of the chemical variables). The full account of the analysis appeared in the journal *Environmetrics* (Reyment, 1996).

Finally, we shall check that we have really done what we started out to do, namely, the decomposition expressed by equation (3.25),

$$\mathbf{X} = \mathbf{G}\mathbf{H}^T.$$

To do this, we have given you a little matrix multiplication program (which you can, of course, use in any other connexion requiring the multiplication of matrices).

Type *multest* at the *DOS*-prompt. You will be asked to supply the name of an input file. Type *multest.chk*. You should be rewarded with the matrix with which you started the biplot computations. What has been done is that $\mathbf{G}$ and $\mathbf{H}^T$ have been multiplied together.

Box 10: Simple example to illustrate the biplot (cf. Gordon, 1981)

Program: **gabriel**

The biplot

Data: **gabriel.dat**

Root = 0.11663

W(j)   −8.78297   0.10833   19.08065   −15.47840

V(T)

$$\begin{bmatrix} -0.0276 & 0.0283 & -0.0003 & -0.0615 & 0.0499 \\ -0.2817 & 0.2886 & -0.0036 & -0.6269 & 0.5086 \\ 0.9346 & -0.9573 & 0.0118 & 2.0798 & -1.6871 \\ -0.1342 & 0.1374 & -0.0017 & -0.2986 & 0.2422 \\ -0.1684 & 0.1725 & -0.0021 & -0.3747 & 0.3040 \end{bmatrix}$$

Canonical weights

columns   −0.008135   0.000587   0.032807   −0.019543

rows   −0.027648   −0.281734   0.934615   −0.134185   −0.168378

data for plotting in file **biplots**

The points arranged for plotting ease

```
                Points for Variables
point   1  0.6177  −0.1048  −0.1391  −0.0127
point   2  0.2511  −0.0764   0.0475   0.0002
point   3  0.4610   0.0717  −0.0128   0.0276
point   4  0.5243   0.1617   0.0379  −0.0224

                Points for Specimens
point   5  0.2364  −0.2651  −0.8334  −0.0223
point   6  0.4220  −0.3211   0.4113  −0.2273
point   7  0.3455   0.0145   0.0888   0.7539
point   8  0.8711  −0.2521   0.0424  −0.1082
point   9  0.4789   0.8621  −0.0922  −0.1358
```

These two matrices, in the order of firstly **H** to which is appended **G** are stored in the file *biplots* for graphical appraisal.

PROGRAM MULTEST

Instructions for using the program **multest**

Line 1: 1   number of rows in the first matrix,
        2   number of columns in the first matrix,
        3   number of columns in the second matrix

Line 2 and following contain the matrices

The "commas" in line 1 – a C-requirement – are necessary here. Note also that the number of columns in the first matrix (here **G**) is the same as the number of rows in the transpose of **H**. You can see how to set up the input file by consulting *multest.chk*.

Practical Note: The graphs yielded by the biplot can be made more useful by a few simple constructions. Find the centre-point and from it draw lines to the points denoting the variables, which should be labelled, of course. Try then to find plausible connections between the locations of the variables and the objects.

SUMMARY

1   Principal Component analysis and the *R*-mode.
2   Principal component factor analysis.
3   Cross validation.
4   Stability of estimates.
5   Log-contrast principal components for compositions.
6   Principal coordinate analysis and the *Q*-mode.
7   Log-contrast principal coordinates for compositional data.
8   Singular value decomposition.
9   Simple analysis of correspondences.
10 The Gabriel biplot.
11 Programs and associated training sets

    pcomp1   pcomp.dat
    pcomp2   afrobol.dat
    pcaident   matinv2.dat
    pcvalid   alkaval.dat
    jknfpca   keyella.dat
    pcaconst   haitipc.dat
    pcoord   afrocrd.dat
    pcrdcons   canary.dat
    singval   matrx3.dat

benzec   fenantrn.dat
gabriel   gabriel.dat
             leptobip.dat
multest   multest.chk

12 **Appendix**: Step-by-step account of how to apply *Graph Server* to a data-set of coordinates in a very simple manner. Please note that this example only makes use of very few of the available functions of *Graph Server* and *Graph Wizard*. Details on how to use these programs are given in chapter 2.

1 Double-click on the file *GraphWizard* (or on its shortcut, if you created one previously).
2 Click on button NEXT and then select BROWSE.
3 Find the data to be processed.
4 Select the file.
5 Click on **Select all** (or manually select the data to process, if the file contains extraneous data you want to discard). Click on NEXT.
6 Select the proper organization of the data, e.g., 'columns contain coordinate values'. Click on NEXT.
7 Click on **add new data-set** For instance, you may want to examine the plot of the scores on the first two axes. Therefore, select columns '1' and '2' and then as a marker, choose, say, NUMBERS. Press NEXT.
8 Adjust the ranges of the coordinates and insert a scale marking. In most cases, you need to adjust the range values. For instance, when Graph Wizard proposes to display the range between 31.20 and 58.47 with tick marks every 9.25 units, you can safely round these values to 30.60 and 10, respectively. You may need to indulge in a sequence of trials in order to arrive at the optimum result for the neatest graph. Press NEXT.
9 Select the appropriate graph-settings. You may need to experiment with different settings at this stage as well.
10 Type in the labels for the ordinate and abscissa, and choose appropriate fonts and font sizes. NEXT.
11 Tick one or more of the boxes if you want to save the graph for later use, or just press FINISH to start Graph Server and display the graph.

Your processed plot will now appear on the screen. A good quality print can then be easily obtained by following the appropriate procedure (see chapter 2).

Chapter 4

Comparing samples from two populations: the discriminant function

INTRODUCTION

It is no more than a matter of practical convenience to give the subject of comparison of samples drawn from two populations a chapter of its own. There is, nonetheless, a certain practical advantage in this since some of the most widely invoked multivariate statistical procedures belong, directly or indirectly, to this category.

One of the earliest multivariate statistical methods to appear was the *Linear Discriminant Function*, devised by R. A. Fisher to provide a quantitative expression of the multivariate difference residing in samples drawn from two populations, the procedure being based on analogy with multiple regression. The original proposal was made in reference to a taxonomic problem, the celebrated case of the species of *Iris*. The method was, however, quickly adopted for a wide range of other classes of data. Actually, Fisher got ahead of himself by helping an Australian doctoral candidate, Miss M. Barnard, to analyse her measurements on Egyptian skulls (Barnard, 1935) by his newly conceived method. This application, though biologically very unsound, is still interesting for people concerned with developing models for linking time-differentiation to discrimination. Discriminant functions are well covered in many statistical textbooks. A good reference is Hand (1981), another is Anderson (1984).

The sample linear discriminant function is computed from the covariance matrix formed by pooling the covariance matrices of each of the samples to yield matrix S_w and the difference d between the respective mean vectors of the samples. The equation for finding the p coefficients f of the linear discriminant function is simply:

$$f = S_w^{-1} d \qquad (4:1)$$

There is a structural similarity to multiple regression here, which is sometimes exploited as a means of providing approximate standard deviations for the coefficients of discrimination by analogy with multiple regression coefficients. A simple example is presented in **Box 11**.

The program *disfun.exe* computes the example in **Box 11**. It also provides the generalized statistical distance corresponding to (4:1), the associated test of significance (the Hotelling T^2), and the discriminant function scores with an assessment of how well the two sets of data have been separated by the function. This method of computing efficiency of the function is a somewhat subjective procedure, however, being based on the assignment of the individuals of each sample on the basis of the discriminator computed from them. In cases where the two covariance matrices are very different, it may be more appropriate to proceed by some method that can absorb such a difference. One such procedure is the *quadratic discriminant function*, another way of treating heterogeneous data is offered by the generalization of a technique of univariate statistics for unequal variances.

Rao (1949; 1952) made several important observations about the workings of the discriminant function and the generalized distance. One of these concerns the value or demerit of adding more and more variables to a study in the hope that separation will be bettered. If the increase in the generalized distance is not appreciable, the reduction in errors of misidentification by adding more characters is negligible. The presentation given here does not embrace more recent specialist results. If you wish to read further, we can recommend the book by Krzanowski and Marriot (1995).

Instructions for using the program **Disfun**

Line 1: The dimensionality of the problem

Line 2: Title of the job

Line 3: 1 size of first sample
        2 size of second sample
        3 1 if extra observational vectors are to be compared
           0 for default (no extra vectors)
        4 1 for input as a data-matrix (organized as below)
           0 if data as pre-computed covariance matrices and means
        5 1 if full output from the program is desired
           0 for just the essential results
        6 1 if logarithms of the input vectors are to be taken
           0 for raw data

[line 3 applies only if item 4 contains a "1": i.e. extra vectors can only be read in the case of the data-matrix option]

Introduction

Line 3: For *matrix input*
          covariance matrices first
          corresponding mean vectors next

or,

Line 3: For input as the *data-matrix*

Line 4: The title of the sample

Line 5+: The sample

The taking of logarithms in Line 6 is perhaps more appropriate for divergent biological data which are to be brought into conformity with the multivariate normal distribution. The discriminant function scores are stored in files **disscore**, for the linear discriminant function, and **quadscor** for the quadratic discriminant function. These files are useful for making histograms by means of *Graph Server*. However, the program yields a set of printer plots of histograms of the discriminant function scores which often prove useful for quick appraisals.

Example of discrimination

A specimen of how to set up the file for computing the discriminant function in *disfun.exe* is given in *echindfn.dat*. This data-set presents measurements on five morphological characters of the carapace for two species of the Eocene ostracod genus *Echinocythereis* (length and height of the shell, anterior height, posterior height, length of the posterior process). Further details are to be found in Reyment (1985).

Some of the output generated by the program is listed in **Box 11**.

Box 11: Example of the discriminant function program **disfun** applied to the data for evolution in species of *Echinocythereis* from the Eocene of Aragon, Spain. Data from Reyment (1985).

Program: **disfun**

Data: **echindfn.dat**

These data consist of a sequential sample (an evolutionary series) of means for *Echinocythereis isabenana* on which five characters are available, to wit, measurements on length and three heights of the carapace, plus the length of the posterior process (the older suite) and the younger suite, for *Echinocythereis posterior*, likewise composed of sequentially ordered means.

The pooled covariance matrix

	1	2	3	4	5
1	0.00211	0.00097	0.00152	0.00115	0.00012
2	0.00097	0.00062	0.00073	0.00054	0.00010
3	0.00152	0.00073	0.00126	0.00087	0.00013
4	0.00115	0.00054	0.00087	0.00086	0.00007
5	0.00012	0.00010	0.00013	0.00007	0.00013

Pooled correlation matrix

	1	2	3	4	5
1	1.0000	0.8472	0.9324	0.8576	0.2372
2	0.8472	1.0000	0.8249	0.7394	0.3526
3	0.9324	0.8249	1.0000	0.8395	0.3100
4	0.8576	0.7394	0.8395	1.0000	0.2137
5	0.2372	0.3526	0.3100	0.2137	1.0000

As is so often the case for crustaceans, some distance-measures are highly integrated. Here, the length and three heights display evidence of high correlation. The posterior process can be seen to be much less strongly bound to those variables.

Pooled standard deviations

0.046   0.025   0.035   0.029   0.011

The mean vectors

vector 1
0.9840   0.5627   0.7509   0.5976   0.1294
vector 2
0.7954   0.4679   0.6132   0.4841   0.1075
difference mean vector
0.1885   0.0948   0.1377   0.1135   0.0219

Introduction

	Disc coeff.	Stand. coeff.	Classical coeff.
1	63.160	2.900	14.715
2	26.058	0.651	6.071
3	−23.242	−0.824	−5.415
4	47.655	1.395	11.103
5	84.073	0.958	19.587

Three ways of expressing discriminant function coefficients are illustrated in the foregoing array. The first column contains the directly computed values, the second column holds the standardized coefficients, and the third column has the classically derived coefficients of Fisher (1936). The efficiency of the discriminant function can be verified by a test based on the generalized statistical distance (the test is usually known as the Hotelling T^2).

$D^2 = 18.42$, $D = 4.29$, $F = 89.34$ with 5 and 96 degrees of freedom. Probability that the samples are from same distribution <0.0001.

probability of misidentification = 0.0159 based on $D/2$

unbiased $D^2 = 17.12$, unbiased $D = 4.1377$

unbiased misidentification probability = 0.0193

Unbiased D^2 is slightly lower than the biased value, which seems to be commonly the case in many investigations.

The value obtained of the generalized statistical distance is highly significant, as indicated by the figure for the variance ratio, computed from the Hotelling T^2. There is, therefore, a low probability of making a wrong assignment of a specimen. This can be gauged approximately by running the original observations through the discriminant function and seeing how the specimens are assigned.

discriminant mean 1   98.711

discriminant mean 2   80.287

Identifications for Sample 1

46 correct, 0 wrong for a percentage of 0.0 % wrong

This line in the output says that all 46 specimens of the older sample were correctly identified by the internal process of passing the data through the discriminant function computed from them. If you have access to more material, and know it to be correctly identified, it would be more to the point to see how it was treated by the computed discriminant function.

Identifications for Sample 2

56 correct, 0 wrong for 0.0 % wrong

Again, for the second sample, all specimens were correctly picked out by the linear discriminant function.

The file *disscor*, generated by the program, can be inserted into *Graph Server* to illustrate how efficient the discriminant function is. This is demonstrated in Fig. 29. The plot shows that the discriminant scores for the two samples are completely separated, thus attesting to the efficiency of the function for distinguishing between the two categories.

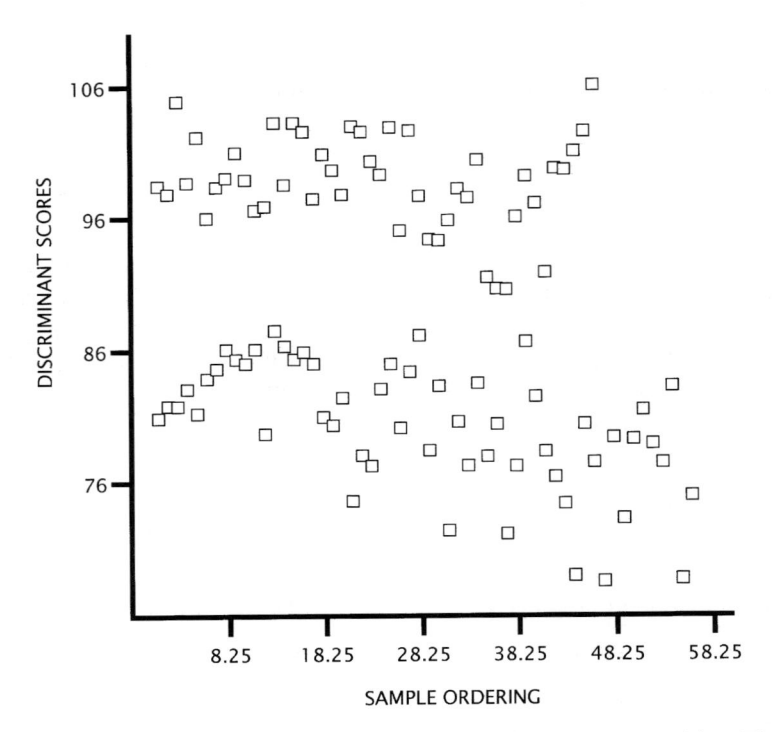

Fig. 29. Plot of the discriminant scores against sample origin for the Eocene ostracod data. The separation of the two species is very good and it may be concluded that the linear discriminant function is efficient.

Quadratic discriminant analysis
The Box test for homogeneity of covariance matrices yields

chi-square = 62.78 for 15 degrees of freedom.

A test for homogeneity of covariance matrices gave the above result, namely, that there is significant heterogeneity in the dispersions. There is therefore motivation for a specific test, which is done by means of the quadratic discriminant function.

Quadratic discriminant scores

sample 1: 2.17% wrong of 46 specimens
sample 2: none wrongly assigned (of 56 specimens)

In this case, the result was not as good in that a specimen of the first sample was incorrectly identified.
In conclusion, it can be claimed that there can be little doubt that the samples are drawn from statistical populations that differ greatly from each other with respect to the characters measured.

QUADRATIC DISCRIMINATION

When the covariance matrices are unequal, and or, the distributions are non-normal, recourse can be made to a quadratic discriminant function. Briefly, a quadratic discriminant function can be constructed as follows (Seber, 1984, p. 297):

$$Qs(\mathbf{x}) = \tfrac{1}{2}(\log(|\mathbf{S}_2|/|\mathbf{S}_1|) - (\mathbf{x} - \bar{\mathbf{x}}_1)^T \mathbf{S}_1^{-1}(\mathbf{x} - \bar{\mathbf{x}}_1) + \mathbf{x} - \bar{\mathbf{x}}_2)^T \mathbf{S}_2^{-1}(\mathbf{x} - \bar{\mathbf{x}}_2)) \qquad (4:2)$$

The effect of applying eqn. (4:2) is exemplified below in **Box 12**.

Box 12: *Example to exemplify quadratic discrimination.*

Terebratella retusa versus *T. septentrionalis*

The traits measured on two species of brachiopods are length, height and width of the shell, two lengths on the foramen, and the weight of the shell, six characters in all (Endo et al., 1995). Firstly a standard linear discriminant function is computed then a quadratic discriminant function, which is compared with the linear discriminant function.

$D^2 = 5.31$, $D = 2.30$
$F = 21.38$ with 7 and 111 degrees of freedom.

The probability that the samples are not from same distribution is less than 0.00001. It is therefore quite clear that the two species are very different on the grounds of the characters determined. However, the probability of wrongly assigning individual specimens is relatively high, as indicated by a direct test as illustrated below.

Probability of misidentification = 0.125 (based on $\frac{1}{2}D$)

$D^2 = 4.71$, $D = 2.17$

misidentification probability = 0.139
Assignment results for Sample 1

mean = 15.38; standard deviation = 2.19

51 specimens correctly assigned, 7 wrong; i.e. 12.1% wrongly assigned

Assignment results for Sample 2

mean = 20.69; standard deviation = 2.40

54 correctly assigned, 7 wrong; i.e. 11.5% wrongly assigned

Quadratic discriminant analysis of the same data set as above

Homogeneity test for matrices S_1 and S_2
chi-square = 79.25 for 28 degrees of freedom

This value is significant and it may be concluded that there is significant heterogeneity in the covariance matrices on the criterion used. However, the data also diverge from multivariate normality and it is therefore interesting to see what the quadratic discrimination method discloses. Briefly, the assignment results are as follows:

sample 1 10.34% wrong of 58 specimens
sample 2 8.20% wrong of 61 specimens

Hence, the quadratic discriminant function brings about a slight improvement in the diagnostic ability of the discrimination process.

Discussion of the quadratic discrimination example

The quadratic discriminant classification in **Box 12** has produced a slight though important improvement in the efficiency of the discrimination. Granted that the covariance matrices are heterogeneous, the improvement could possibly also be due to non-linearity in the data. The method of quadratic discrimination provides a means of proceeding when the covariance matrices are heterogeneous. It is only safe to use it when large samples are available, at least 50 specimens in each of both groups. If one variate has a very small variance and, or, several variates are highly correlated, spurious discrimination can sometimes result.

THE GENERALIZED DISTANCE AND HETEROGENEOUS DISPERSIONS

We shall now introduce a program that is more explicitly designed to treat data that differ in the configuration of the sample hyperellipsoids. For most purposes, *disfun.exe* is perfectly adequate; however, where interest lies with dissecting the anatomy of the differences between samples, the present procedure is recommended. Note, however, that it is solely structured for dealing with linearly related variables. The program *het.exe* provides, as an initial step, a univariate appraisal of the data, including tests of univariate skewness and kurtosis. The statistical method utilized in the program is based on the Anderson–Bahadur T^2 for heterogeneous covariance matrices, from which a corresponding generalized distance can be readily obtained as well as the appropriate discriminant function (Anderson and Bahadur, 1962).

Instructions for using the program **Het**

The data are entered in free format.

Line 1: 1 Number of variables
    2 Size of the larger sample
    3 Size of the smaller sample
        1 specifies pre-processed data
    4 1 specifies that logarithms are to be taken

0 is for raw observations
5   1 specifies that the usual discriminant function is to be computed no matter what the program decides is the better solution.
Default is 0

Line 2: The title of the job.

Line 3: The data in the form specified by item 4 on line 2.

The specifications 2 and 3 need only be applied if sample sizes are very different (for geometrical reasons or stability with respect to orientations of distributions) for the comparisons). In the ensuing example, the sample sizes are of roughly the same order of magnitude. An abridged listing of the results obtained by inserting the file *brachhet.dat* into program *het* now follows. These data are measurements on brachiopod shells of the genus *Terebratella* (cf. Endo et al., 1995). It is interesting to see what a specifically constructed analysis of the geometry of the empirical distributions can disclose. The data of *brachhet.dat* are logarithmically transformed. The main results of the computations are reproduced in **Box 13**.

Box 13: Example of discrimination and generalized distances for unequal covariance matrices using data on two species of brachiopods of the genus *Terebratella* (Endo et al., 1995).

Program: **het**

Data: **brachhet.dat**

T. retusa versus *T. septentrionalis*

The characters measured are length of the ventral valve (1), width of the shell (2), height of the shell (3), density of ribbing (4), length of the foramen (5), width of the foramen (6) and weight of the shell (7). The full treatment of the data is in Endo et al. (1995). The reason for doing the analysis was to see how well the standard measurements could distinguish between species established on purely biological grounds.

"reference" sample size is 61 specimens

comparison sample size is 23 specimens

Distribution details for reference sample

Variable	Skewness	t-value	Kurtosis	t-value
1	−0.1739	−0.5679	0.1983	0.32
2	−0.4455	−1.4546	0.4955	0.82
3	0.0987	0.3221	0.3400	0.56
4	−1.2171	−3.9738	2.3354	3.86
5	−0.3634	−1.1865	−0.6391	−1.05
6	−0.4223	−1.3789	0.0274	−0.04
7	0.8553	2.7927	0.1691	0.28

Distribution details for comparison sample

Variable	Skewness	t-value	Kurtosis	t-value
1	0.0318	0.0660	−0.8836	−0.94
2	0.3663	0.7610	−0.7418	−0.79
3	0.7305	1.5177	−0.1174	−0.12
4	−0.1688	−0.3506	−1.2563	−1.34
5	1.0052	2.0883	1.1591	1.23
6	0.6692	1.3902	−0.0629	−0.06
7	0.5933	1.2326	−0.7978	−0.85

These univariate tests show that most variables are univariate normally distributed. Variable 4 for the first sample deviates both with respect to skewness as to kurtosis. For the second sample, variable 5 is significantly skewed at the 5% level.

The program outputs covariance matrices and their connected latent roots and vectors for purposes of analysing the orientations of the dispersion ellipsoids in relation to each other. The following results pertain to the non-central chi-square distribution and a specific test for heterogeneity in covariances, and for which a special set of tabulated values is required for their interpretation. It is, however, an easy matter to transform the non-central chi-square parameters to standard values of chi-square; this is carried out automatically by the program.

$B^2 = 73.19$

$\beta^2 = 3.44$

These data transform to a standard chi-square of 64.20, which for 28 degrees of freedom is highly significant (the tabulated value is 41.33). The conclusion to be drawn here is that the covariance matrices of the samples for the two species of brachiopods are indeed very different from each other. We shall now see in what manner these differences are manifested.

As noted, the foregoing result points to there being strong heterogeneity in covariances. We shall test whether this can be ascribed to differing orientations of the hyperellipsoidal axes. Note that in the following, a χ^2_6 value of 12.59 indicates significance at the 5% level. There are only three axes of interesting length; the remaining four were rejected by the decision process in *het.exe*.

Orientations of ellipsoidal axes

vector	chi-square	df
1	16.77	6
2	47.21	6
3	44.06	6

The test of significance for these three axes indicates them to be significantly rotated in relation to each other. This indicates at least one source to the heterogeneity in dispersions (Anderson, 1963; Reyment, 1969).

Statistical distance for heterogeneous covariances

The following discriminant vector (i.e. the coefficients of the linear discriminant function) was obtained by the Anderson-Bahadur iterative procedure.

Estimate of discriminant vector
1	−3.60
2	2.83
3	0.24
4	−0.11
5	8.28
6	−21.40
7	27.77

Most of the discriminatory power lies with variables 5, 6 and 7. (The program outputs a histogram of the discriminant scores. This is not reported here.)

The generalized distance and heterogeneous dispersions

Chernoff's separation T-criterion

In probing the properties of a generalized distance, it can be useful to determine how much of the distance is due to differences in the centroids and how much is to be put down to differences in the covariance matrices. A useful (though little known) partitioning procedure is that of Chernoff (1973), which is included in the calculations performed by *het.exe*. Note, please, that this T is not the same as the generalized Student's t of Hotelling. Hand (1981) gives an illuminating account of the use and interpretation of the Chernoff criterion.

distance-separation due to means = 2.08
distance-separation due to unequal covariances = 3.11
Chernoff's separation $Tc^2 = 5.19$

This analysis suggests that a somewhat greater part of the statistical distance between the two populations of brachiopods is due to inequality in covariances.

Results for Mahalanobis D^2

D	D^2
2.60	6.76

significance for D^2 and T^2

$T^2 = 112.93$
$F_{7,76} = 14.95$

Discussion of the analysis in **Box 13**

The covariance matrices have been shown to differ in their geometrical properties, which means that the computations for producing a generalized distance and accompanying discriminant function should proceed via some appropriate method, such as the Anderson–Bahadur generalization of the Behrens–Fisher problem (one of the classical problems of mathematical statistics) for heterogeneous variances in the univariate analysis of variance. It is a moot point whether this step can always be justified in practical situations since the improvement in the result is usually rather slight. It is for this reason, we have, as an exploratory measure, made the program *het* include a standard calculation of the generalized distance in order to provide a means of comparing how useful the Anderson–Bahadur step really is for a particular problem. You can read more about the theory underlying the test in Anderson (1984). The details of the Chernoff procedure can be consulted in Chernoff (1973). Briefly, Chernoff's criterion represents the sum of two terms. One may be regarded mainly as a Mahalanobis distance corresponding to a weighted average of the two covariance matrices. This term gives an expression for the distance between the means. The second term is mainly a measure of the information contributed by the differences between the two covariance matrices.

DISCRIMINANT ANALYSIS FOR COMPOSITIONS

Aitchison (1986) demonstrated that the usual arsenal of multivariate methods is available for compositional data with but few modifications. For most purposes, the sample log-ratio covariance matrices $S_{(i)dxd}$ are suitable vehicles for computation of a linear discriminant function, following the steps given above. As you already have been told, the log-ratio covariances entail a loss of one part and the dimensionality of the discrimination problem becomes $D - 1 = d$. The computation of a compositional Mahalanobis generalized statistical distance follows on directly. The advantage of using the log-ratio covariance matrices lies with their being positive definite and can, therefore, be inverted by standard procedures. The *generalized likelihood ratio test* requires a little more effort in order to make its use valid for compositional data. The test can be constructed in order to provide an answer to either of two questions. The first of these is the test:

$$\chi^2_{df} = N_1 \log(|S_c|/|S_1|) + N_2 \log(|S_c|/|S_2|) \tag{4:3}$$

The terms in (4:3) are defined as follows. If S_p denotes the pooled sample covariance matrices, and $\mathbf{m}_{(i)}$ ($i = 1,2$) the appropriate compositional sample mean vectors, then,

$$S_c = S_p + (N_1 + N_2)^{-2} N_1 N_2 (\mathbf{m}_1 - \mathbf{m}_2)(\mathbf{m}_1 - \mathbf{m}_2)^T. \tag{4:4}$$

Equality of means and covariance matrices is tested on degrees of freedom $\frac{1}{2}d(d + 3)$, where d is $D - 1$. The second construct is for testing equality of covariance matrices alone; here, df $= \frac{1}{2}d(d + 1)$ and the appropriate chi-square formulation is:

$$\chi^2 = N_1 \log(|S_p|/|S_1|) + N_2 \log(|S_p|/|S_2|). \tag{4:5}$$

When the covariance matrices are significantly different, the standard procedure for the multivariate Behrens–Fisher solution is available (see procedure presented in the foregoing example in **Box 13**).

A major preoccupation of petrologists is assigning specimens to a priori established rock-types. The original semi-quantitative methods of classical petrology are gradually being ousted by automated procedures. Crude linear discriminant functions have been widely applied to allocating specimens (LeMaïtre, 1982), a procedure that is clearly open to critical evaluation, but which is still being used in petrological work with the justification that any transformation (such as is engendered by the computation of the log-ratio) of data must necessarily distance the analyst from the problem and, hence, Science. This is a further case of the existence of the need for a broad and well balanced information campaign in the field of statistical geochemistry.

The "training set technique" for establishing allocation criteria in discriminant analysis introduces *resubstitutional bias*, as is well known to statisticians, and as has already been mentioned (see also Krzanowski and Marriot, 1995).

The linear discriminant function, as presented in the foregoing section, is available for compositional data in two forms in that either the log-ratio covariance matrix can be used, or the centred log-ratio covariance matrix. In the latter case, a pseudo-matrix-inversion will be necessary, for example, by invoking the Moore–Penrose method, which exploits the spectral decomposition:

$$\Gamma^- = \lambda_1^{-1}\mathbf{a}_1\mathbf{a}_1^T + \ldots \lambda_d^{-1}\mathbf{a}_d\mathbf{a}_d^T + 0\mathbf{a}_D\mathbf{a}_D^T$$

The λ_i are the latent roots obtained from the covariance matrix by any standard method and the $\mathbf{a}_i$ are the associated latent vectors. There can, of course, only be d non-zero latent roots, owing to the constraint.

The program *dfnconst.exe*, which is essentially an appropriately adjusted version of program *disfun.exe*, uses the log-ratio covariance matrix for producing a simple discriminant function analysis; the calculations also include the corresponding quadratic discriminant function. An example is given in **Box 14**. This treats the data for chemical determinations on the shell of a species of the ostracod *Leptocythere* observed at two stations, Roscoff and in the Baltic Sea (Bodergat et al., 1993; Reyment, 1996) in an environmetric study. The pertinent data-file is called *leptodfn.dat*. There are 12 elements to consider, but the problem reduces to one of 11 dimensions since the one part enters into the common divisor used for producing the log-ratio covariance matrices. In a foregoing section it was pointed out that the usual method of computing linear discriminant functions will not work with compositional data because of the problem occasioned by the inversion of a singular covariance matrix. This difficulty is obviated by the use of the log-ratio transformation (Aitchison, 1986).

Instructions for using the program **dfnconst**

Line 1: The dimensionality required (i.e. the number of parts).

Line 2: 1 size of first sample
         2 size of second sample
         3 1 for full output; 0 is default (reduced output)

Line 3: Title of the job

The data sets now follow, each preceded by its title.

Output details

The use of the log-ratio covariance matrix requires division by one of the parts (where relevant, usually SiO_2). As already noted earlier on, this manipulation reduces the dimensionality of the problem by one. The program computes basic statistics for each of the samples. It provides the coefficients of the linear discriminant function in three slightly different variants, depending on adjustments for the variability of parts and "normalization". The Mahalanobis generalized distance and the associated test for significance are computed, both for biassed and unbiassed D^2.

Plottable data are stored in *dfnscore* for the linear discriminant scores, and *quaddfn* for the quadratic discriminant scores. These files can be used for making histograms of the separation.

Box 14: Example of compositional discriminant analysis.

Program: **dfnconst**

Data: **leptodfn.dat**

Linear discriminants for compositions

Leptocythere psammophila sampling stations: elements determined are: Ca Ba Cl S Sr Fe Mn Na Mg Al Si P, 12 in all. The common log-ratio divisor is Si and there are 12 parts, but the dimensionality of the analysis is 11.

Discriminant function for compositional data

	Disc coeff.	Stand. coeff.	Classical coeff
1	37.077	4.352	8.434
2	12.207	2.713	2.777
3	2.637	1.684	0.600
4	21.075	5.207	4.794
5	−52.130	−7.891	−11.858
6	1.467	0.872	0.334
7	3.348	1.274	0.762
8	0.910	0.432	0.207
9	−27.403	−5.477	−6.233
10	−0.866	−0.542	−0.197
11	−2.901	−2.853	−0.660

Not unexpectedly, most of the discriminatory power lies with parts Ca, S, Sr and Mg.

$D^2 = 19.33$ and $D = 4.40$
$F = 7.57$ with 11 and 16 degrees of freedom.

The probability that samples are from same distribution is 0.0002. There is therefore a clearly manifested difference between the two samples.

Result for first sample

14 correctly allocated by function, 0 wrongly assigned.

Result for second sample

14 correctly allocated by function, 0 wrongly assigned.

Total $N = 28$, wrongly assigned $= 0\%$.

Quadratic discriminant analysis

The Box test of homogeneity for covariance matrices

chi-square $= 178.81$ for 66 degrees of freedom

which is significant at the 5% level at least. This is, however, not serious enough to influence the efficiency of the linear discriminant function, as is exposed by the results for the quadratic calculations listed below.

sample 1   0 percent wrong of 14
sample 2   0 percent wrong of 14

In the present example, the log-linear discriminant function succeeds in providing efficient discrimination between samples. Remember, however, that the example recycles the input data and this, you will recall, tends to give an over-optimistic view of the efficacy of the computed function for making correct assignments. You will also see that the quadratic discriminant function is an unnecessary codicil to the analytical protocol in that it made little improvement to the discrimination.

Discriminant analysis of spectroscopic data

We make here brief mention of the use of cross-validation in discrimination, but do not provide a program for doing the calculations. Mertens (1998) has presented a combination of techniques, based on cross validation principal component analysis and discrimination for the analysis of spectroscopic data. Such data are usually high-dimensional for which very few samples are available. The procedure advocated by Mertens uses exact principal component influence theory and a new class of influence measures based on ratios of Euclidean distances in orthogonal spaces.

SUMMARY

1 Linear discriminant function
2 Quadratic discriminant function
3 Generalized statistical distance and heterogeneous dispersions
4 Chernoff's separation criterion
5 Discriminant analysis for compositions
6 Programs and associated training sets
   disfun   echindfn.dat
   het   brachhet.dat
   dfnconst   leptodfn.dat

Chapter 5

Analysis of several groups: canonical variate analysis

INTRODUCTION

If you want to analyse several groups of observations, that is, samples drawn from more than two statistical populations, then the method of canonical variates is an appropriate choice. In some respects, it can be represented as being a generalization of the linear discriminant function, just reviewed, in others it can be thought of as a kind of generalized principal components model. It has, of course, its deficiencies, as do many multivariate methods. These drawbacks are mainly concerned with:

1 The fact that the information supplied is in the form of a priori delineated groups (that is, groups recognized by the analyst), which creates a tendency to reinforce the segregation already implied by the structured nature of the input. This is not a fatal fault and if there are really undecided specimens in the material, they are usually efficiently disclosed and correctly allocated to the appropriate group.

2 It is a popular exercise to attempt to reify the canonical vectors by analogy with widely (though not generally) accepted procedure often employed in connection with principal component analysis. This can only done with extreme caution owing to the fact that the components of the canonical vectors tend to be unstable under repeated sampling. This condition is particularly noticeable when there are high between-groups correlations, such as occur in morphometrical work. Campbell (1979, 1980) has devoted special attention to aspects of stability in applied canonical variate analysis. A useful summary of Campbell's results is given by Seber (1984). There is a related condition of stability of coefficients in multiple regression analysis, a fact that seems to have been first noticed by Campbell (1980), in detail at least.

In algebraic terms, the first canonical variate is that linear combination which maximizes the ratio of between-groups sums of squares to the within-groups sums of squares for a one-way multivariate analysis of variance of the canonical variate scores. For k groups and p variables we have the canonical variate scores

$$y_{ij} = \mathbf{C}^T x_{ij} \qquad (5:1)$$

where x_{ij} denotes the i-th of N observations for the j-th group. The first canonical vector is derived so as to maximize the ratio

$$f = \mathbf{c}^T \mathbf{B} \mathbf{c} / \mathbf{c}^T \mathbf{W} \mathbf{c} \qquad (5:2)$$

where $\mathbf{B}$ is the between-groups matrix of sums of squares and cross products and $\mathbf{W}$ is the within-groups matrix of sums of squares and cross products. These matrices are formed as follows:

1 Compute $\mathbf{T}$, the matrix of sums and squares and cross products for all the groups pooled into a single data-matrix.
2 Compute $\mathbf{W}$ by adding together the matrix of sums of squares and cross products for each group.
3 Find then the difference $\mathbf{B} = \mathbf{T} - \mathbf{W}$.

The canonical vectors $\mathbf{c}$ and the canonical roots f satisfy eqns (5:3) and (5:4) below:

$$(\mathbf{B} - f\mathbf{W})\mathbf{c} = \mathbf{0} \qquad (5:3)$$

and

$$|\mathbf{B} - f\mathbf{W}| = 0 \qquad (5:4)$$

The canonical vectors are usually scaled so that

$$\mathbf{c}^T \mathbf{W} \mathbf{c} = N_w \qquad (5:5)$$

where N_w is the within-groups degrees of freedom.

There are $\min(k - 1, p)$ non-zero canonical roots to the solution of the determinantal eqn. (5:4). If there is closure in the data, there may be less than p non-zero roots when $p < k$. This topic is taken up on p. 182. The expression "min" says that whichever is smaller, $k - 1$ or p, indicates the number of non-zero canonical roots. The program for canonical variate analysis is an updated version of the one originally published in Blackith and Reyment (1971). In addition to the steps outlined above, the following computations are done by the program **cva.exe**.

1 Generalized statistical distances between all pairs of groups together with the relevant tests of statistical significance. These are the Hotelling T^2-values, transformed to the appropriate variance ratios.

2   Q–Q-probability plotting for each group, if asked for. This can be a useful step in exploratory work in that it provides a graphical means of assessing one aspect of multivariate normality in a sample. Although a serviceable tool, Q–Q probability plots are no more than advisory in nature and they should always be accompanied by a more comprehensive analysis before it is decided to suppress observations in the interests of approximating the multivariate Gaussian condition. These plots do not actually **prove** anything. The plot is made using the generalized statistical distances computed for each specimen in the sample. If there are strongly divergent specimens in the data, these will show up as isolated points at the top of the graph. The data for doing a Q–Q probability plot are stored in the file *qdist*. The file is composed of sets of coordinates, one for each of the groups, with the pertinent identifications. Q–Q stands for Q(uantile)–Q(uantile) A general account of the concept of probability plotting is given in Hoaglin et al. (1985, pp. 432–441). Briefly, in order to make a graphical comparison between two distributions, or between two samples, or between a sample and a distribution, the quantiles of the one are plotted against the corresponding quantiles of the other. When the two distributions are exactly the same, the plotted points will lie on a straight line with a slope of 1 and which passes through the origin for many but not for distributions that have the same shape but which differ in location and scale (for example,the Gaussian distributions). Deviations show up as (a) outlying points at the 'top' of the plot and sinuosities in the 'line'. The method of calculation programmed here is that of Ramberg and Schmeiser (1972) and used in a geological connexion by Campbell and Reyment (1980).

3   Basic statistical summary for each group, if asked for. This comprises means, standard deviations, both with their confidence intervals, and the covariance and correlation matrices.

4   The latent roots and vectors of the covariance and correlation matrices for each sample, if asked for.

5   Scores for producing selected plots on canonical variate axes. These are saved in two files. The file *canplot1* contains the 'raw' transformations, that is the transformed values unadjusted for the variances. The file *canplot2* contains the same set of values but standardized to unit variance.

6   The minimum spanning tree (often associated with the name of Prim, an early worker in the field) between sample centroids. The centroids (multivariate mean values) are saved in the file **canmeans**. The table of linkages displayed in the output can be drawn into the diagram produced by *Graph Server* by means of some such facility as **MSPAINT** or **CORELDRAW**.

7   A one-way analysis of variance.

8   A choice of performing the computations in either the space of the covariances or that of the correlations. The latter has the often-beneficial effect of standardizing the variances and transforming the original data to spheroidal rather than ellipsoidal distributions. The spherical transformation is often successful for achieving a stable analysis and for that reason is widely used in Campbell's

imaginative contributions to multivariate statistical analysis. The correlational alternative produces a file of values named *cam* that are in a format suitable for use with **shrinkcv.exe**.

Input Information for using **cva**
Line 1: The title of the job

Line 2: 1 number of variables
        2 number of groups
        3 0 for input as 'processed' matrices
           1 for input as data matrices
        4 If the computations are to be made in the space of the covariances, type 0, for the space of the correlations, type 1
        5 1 for Q–Q-probability plots

The data-matrices for each sample now follow in free format.
Ahead of each of the data-matrices, the following information is to be provided:

Line 1: 1 sample size
        2 1 if data to be log-transformed
           0 if not (i.e. "raw" data to be used).
        3 1 for full output (see below)
        4 1 if the principal component reductions of the covariance and correlation matrices are wanted, otherwise put a zero here.
        5 A scale-factor may be needed. This may be required for reducing the data to standard form, such as microns, as is the case for measurements on microfossils made under the microscope to some micrometer scale or other. Some processors baulk at a non-entry in free format, so it this happens and you do not have a scale-factor to take into account, type a 1 or a 0 here.

Line 2: The name of the group.

Line 3, and following lines, the data-matrix in its usual form (rows by columns).

The trial data are in the file *gastcva.dat*. These data consist of a set of observations on three species of the North American mid-Cretaceous ammonite genus, *Neogastroplites*, to wit, four derived shape expressions for the shell, using Bookstein's (1991) method of relative warps, and counts of rib-densities. The three species are *N. americanus, N. cornutus,* and *N. muelleri*, all described in a publication by Reeside and Cobban (1960) and Reyment and Kennedy (1998). Admittedly, the data-set is not very appealing, but it does represent the type of material often con-

Introduction

fronting the geologist. The primary aim of the original analysis was to establish whether the specific assignment on classical palaeontological grounds could be upheld by morphometric analysis, granted that all three species are highly polymorphic with the appearance of varieties that, superficially at least, seem to be identical in all three taxa.

Output information

The salient features of the analysis of the ammonite data are presented in **Box 15**.

Box 15: Canonical variate analysis of three species of the ammonite genus *Neogastroplites* from the Cretaceous (Cenomanian) of the United States and Canada.

Program: **cva**

Data: **gastcva.dat**

The variables consist of four expressions of shape-variability, obtained by the methods of Bookstein (1991) and counts on rib densities. The shape-descriptors are denoted as sh_1. sh_1, etc.

Shape descriptors for 27 ammonite specimens with rib-frequencies

The correlation space was chosen here owing to the marked difference in properties of the shape expressions as opposed to the rib-counts. The correlation format reduces the variances to a common basis. The shape indicators were obtained by the method of relative warps of Bookstein (1991), being based on points (landmarks) located at four diagnostic sites across the diameter of the shell. Granted that the stability of canonical variates is sensitive to deviations from multivariate normality, the program encompasses Q–Q-probability plotting for each component sample. Example for the first and third samples is illustrated in Fig. 30(a) and 30(b).

Canonical Variate Analysis for Groups = 3
Number of variables = 5

178 *Analysis of several groups: canonical variate analysis*

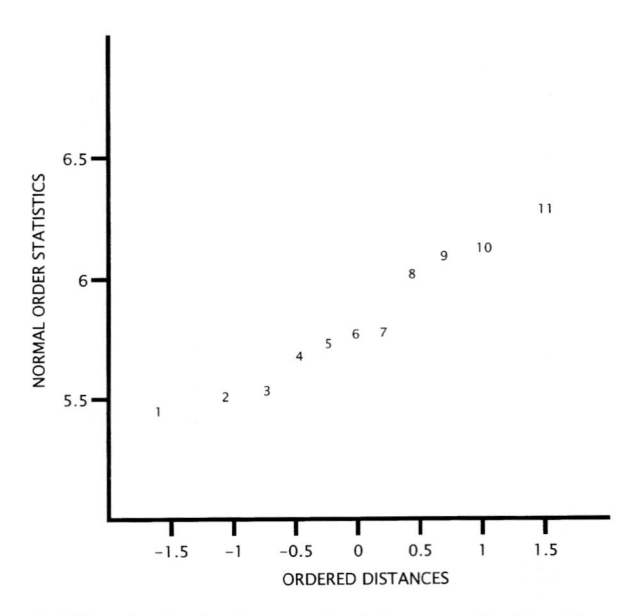

Fig. 30(a). *Q–Q*-probability plot for the first sample of the ammonite data. There are no markedly divergent specimens and hence no indication of deviation from multivariate normality in this sample. The points lie reasonably close to a straight line.

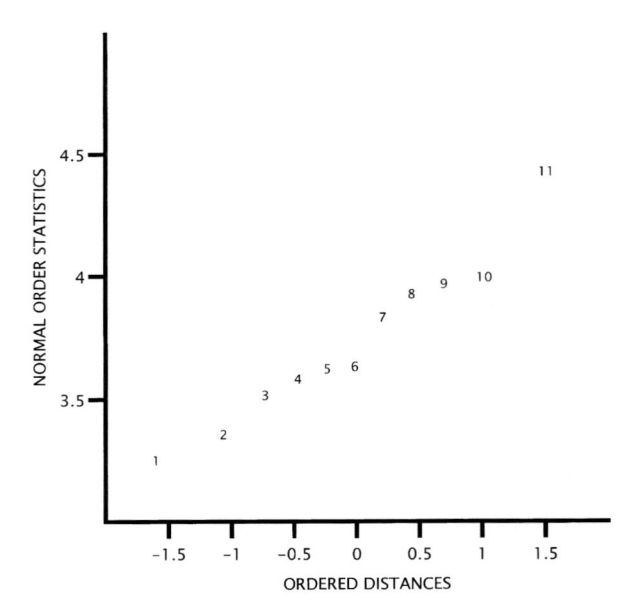

Fig. 30(b). *Q–Q*-probability plot for the third sample of the ammonite data. The first 10 points are approximately linearly located. Point 11 is an obvious outlier.

Introduction

Sample Sizes and Mean Vectors

		Sh_1	Sh_2	Sh_3	Sh_4	Ribs/whorl
1	11.	0.000	0.000	0.001	0.002	26.636
2	5.	−0.001	−0.001	−0.001	−0.010	21.000
3	11.	0.001	0.001	0.001	0.002	21.636

Within-groups matrix standardized to correlations

$$\begin{bmatrix} 1.00000 & -0.14359 & -0.24583 & -0.40210 & 0.39033 \\ -0.14359 & 1.00000 & -0.28893 & 0.21614 & 0.43688 \\ -0.24583 & -0.28893 & 1.00000 & -0.31328 & -0.26212 \\ -0.40210 & 0.21614 & -0.31328 & 1.00000 & -0.03954 \\ 0.39033 & 0.43688 & -0.26212 & -0.03954 & 1.00000 \end{bmatrix}$$

Between groups matrix standardized to correlations

$$\begin{bmatrix} 0.04154 & 0.05318 & 0.01929 & 0.14646 & -0.02982 \\ 0.05318 & 0.07040 & 0.02644 & 0.15379 & -0.06756 \\ 0.01929 & 0.02644 & 0.01028 & 0.04256 & -0.03602 \\ 0.14646 & 0.15379 & 0.04256 & 1.00658 & 0.32199 \\ -0.02982 & -0.06756 & -0.03602 & 0.32199 & 0.39366 \end{bmatrix}$$

The ANOVA indicates that the first three shape indicators do not differ significantly from species to species ($P > 60\%$ for all), whereas the fourth indicator is greatly different ($P = 0.003$). The difference in means of rib-counts is likewise significant, with $P = 0.18$).

Generalized statistical distances

D^2 above diagonal, D below:

	1	2	3
1	0.000	11.929	4.371
2	3.454	0.000	12.916
3	2.091	3.594	0.000

The generalized distances between species 1 and 2 and 2 and 3 are much greater than that between species 1 and 3.

Canonical Root 1 1.9334
Canonical Root 2 0.9915

Canonical Vectors of $\mathbf{W}^{-1}\mathbf{B}$

	1	2
1	0.783	0.775
2	0.170	0.903
3	0.660	0.212
4	1.199	0.012
5	0.012	1.224

Normalized Canonical Vectors

	1	2
1	0.494	0.450
2	0.107	0.525
3	0.416	0.123
4	0.756	0.007
5	0.008	0.712

In very approximate terms, one may claim that the first canonical variate is representative of relationships in shape-indicators 1, 3 and 4. The density of the ribbing is without significance here. The second canonical variate expresses relationships between two shape-indicators, 1 and 2, and the density of the ribbing. This is as far as one dares venture as regards *reification* of the canonical vectors.

Canonical means for axes one and two

0.041	−3.77
−0.046	3.13
−0.020	2.35

Coordinates for plotting canonical means are stored in file *canmeans*, generated by the program.

Significance of latent roots

Canonical Root 1 = 1.765
Chi-square = 38.83 for 10 degrees of freedom

Probability < 0.00001

Canonical Root 2 = 0.689
Chi-square = 15.16 for 4 degrees of freedom

Introduction

Probability = 0.0046.

Both canonical variates are statistically significant, as indicated by the foregoing specific test. This result is important for assessing the relationship between ribbing and shape.

MANOVA test of equality of means

Wilks Lambda = 0.1712
Chi-square = 31.77 with degrees of freedom = 10 and probability = 0.0005.

Here, we have a clear indication of the separation of the three species with respect to the characters being analysed. The canonical variate scores for up to three canonical roots are stored in the file *canplot1*. Fig. 31 illustrates the plot of the scores on the first two canonical variate axes.

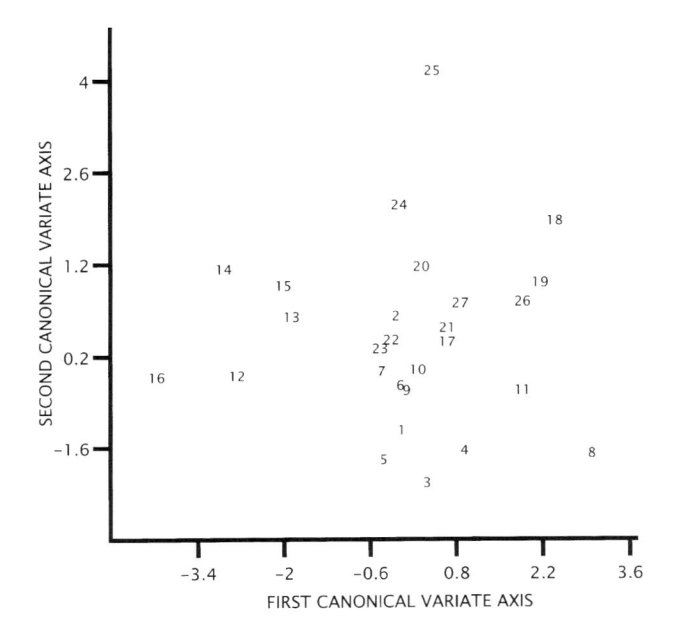

Fig. 31. *Graph Server*-generated plot of the first two canonical variate scores for the ammonite data. There is a tendency for the three species to be segregated into their respective natural groups. Note that specimens 1 to 11 are concentrated in the lower right section of the plot, specimens 12 to 16 to its left, and specimens 17 to 27 in the upper right section.

Conclusions: The canonical variate analysis shows that the species are well differentiated with respect to shape as expressed by the geometric morphometric criteria. *N. americanus* appears to be the more clearly divergent form. The full analysis is given in Reyment and Kennedy (1998).

CANONICAL VARIATE ANALYSIS OF COMPOSITIONS

Aitchison (1986) has shown that the standard model for 'usual' data can be applied without hindrance to compositions. The computations are preferably based on the log-ratio covariance relationships. Log-contrast canonical variates are, as to be expected, uncorrelated. In contrast with compositional principal components, in compositional canonical variate analysis, the centred log-ratio covariance Γ specification has no advantage over the log-ratio covariance matrix, Σ. The modification of our program for computing constrained canonical variates is called *cvaconst.exe*. The common divisor is taken to be the last entry in the vector of parts. Aitchison (1986) showed that it is immaterial which of the parts is selected for the division, unless there is a wish to reify the canonical vectors, which is a procedure of doubtful validity at best (Campbell, 1979).

Input information for **cvaconst**

Line: 1 The title of the job.

Line: 2
     1 number of parts,
     2 number of groups
     3 type 1 for Q–Q-probability plotting, otherwise 0

Line: 3 Heading each of the groups
     1 size of the sample,
     2 principal component analysis requested = 1, otherwise = 0.

  Title of the sample on a separate line

Output details

The canonical variates for one less than the number of input parts, canonical roots, Mahalanobis generalized distances with associated tests of significance, the data for Q–Q-probability plots, stored in the file *aitqdist*, the canonical variate means, stored in *aitmeans*, the unstandardized transformed observations, stored in *aitcval* and the standardized transformed observations stored in *aitcva2*. The appropriate analysis for constrained data-sets follows on in the same manner as is done for log-contrast principal component analysis. Let us look at an example of canonical variate analysis. Compositional data observed on alkaline rocks sampled from mid-oceanic ridges are used to illustrate the procedure. The specimens come from the Middle and South Atlantic ridges and a Pacific Ocean ridge (Sørensen, 1974). The oxides of the elements determined are of Si(1), Ti(2), Al(3), Fe^{+++}(4), Fe^{++}(5), Mn(6), Mg(7), Ca(8), Na(9), K(10), water(11), and P(12), in that order. The analysis was made on the raw data, that is, without accounting for the constraint. The data are located in the file *alkalicv.dat* and a summary of the results is presented in **Box 16**. Although we start with 12 parts, the requirements of the log-ratio transformation necessitate the "loss" of one part and hence a reduction in the dimensionality of the problem to 11 parts.

Box 16: An example of canonical variate analysis for compositional data data on alkaline rocks from the Atlantic Ocean (Sørensen, 1974).

Program: **cvaconst**

Data: **alkalicv.dat**

Canonical Variate Analysis for Groups = 3
Number of parts = 12 (specified above in the main text)

Output for generalized latent roots and vectors

Canonical Root   1   2.272
Canonical Root   2   0.483

Canonical Vectors of $W^{-1}B$

	1	2
1	−0.360	−0.277
2	−0.038	0.000
3	−0.314	0.570
4	0.235	0.154
5	−0.975	−0.068
6	0.488	−0.033
7	0.100	−0.109
8	0.714	0.152
9	0.134	−0.490
10	0.062	0.288
11	0.082	−0.035

Normalized Canonical Vectors

	1	2
1	−0.254	−0.312
2	−0.027	0.000
3	−0.221	0.641
4	0.165	0.173
5	−0.685	−0.076
6	0.343	−0.037
7	0.070	−0.122
8	0.502	0.171
9	0.095	−0.551
10	0.044	0.323
11	0.058	−0.040

The coordinates for plotting the canonical variate scores in *Graph Server* are in files *aitcva1* and *aitcva2*, these being the values for the unscaled and scaled canonical variate axes respectively. If you are interested in safeguarding geometrically relevant relationships in the plot, then the scaled axes are to be selected. For purposes of ordination of points without particular regard to the geometry, then the unscaled axes can often give a more informative diagram.

Significance of latent roots

Canonical Root 1 = 1.579
Chi-square = 56.07 for 24 degrees of freedom

Probability = 0.0002

Canonical Root 2 = 0.394
Chi-square = 13.99 for 11 degrees of freedom

Probability = 0.2327

Manova test of equality of means

Wilks Lambda = 0.206

Chi-square = 41.067 with degrees of freedom = 22, Probability = 0.0082

DISCUSSION OF THE CANONICAL VARIATE EXAMPLE

The alkaline rock data

The results listed above are an abridged account of what is provided by the program *cvaconst.exe*. A few comments will help to guide you through the multitude of details in the output.

The ANOVA, the one-way analysis of variance for each part on four samples, shows that only the mean of one component differ significantly, to wit, potassium (10).

The first canonical root dominates completely in magnitude over the other one (note that the rank of the problem is 2, one less than the number of groups and there are, therefore, only two valid canonical roots). If we confine attention to the first canonical vector (those listed for $\mathbf{W}^{-1}\mathbf{B}$), it will be seen that components (1), (5), (6) and (8) are influential. There is a rule of thumb (cf. Reyment and Jöreskog, 1993) that says that the vector component, when squared, should be 0.1, or greater. (N.B. the elements of a vector are referred to as its "components".) The second canonical vector contains only two influential parts, namely, components (3) and (9).

The significance of the canonical roots is tested. In our present example, only the first canonical root passes the test for statistical significance. The last part of the output is concerned with testing for homogeneity in multivariate means. In the example, the MANOVA (acronym for the multivariate one-way analysis of variance) indicates all mean vectors to differ highly significantly.

A CASE-HISTORY FOR CANONICAL VARIATES: A STUDY IN ENVIRONMENTAL CHEMISTRY

Chemical composition of the shell of the ostracod species, **Leptocythere psammophila** *from Northern Europe*

Bodergat, Carbonnel, Rio and Keyser (1993) studied the influence of environmental chemistry on the composition of the shell secreted by *Leptocythere psammophila* (= the sand-loving leptocytherid). To this end, 41 live individuals were collected from the Baltic Sea (Kieler Förde), North Sea (Sahlenburg/Cuxhaven) and the English Channel (Roscoff). The Baltic locality is one complicated by the possible effects of industrial pollution. The sampling was carried out in Spring (April, 1987), Summer (September, 1988) and Winter (December, 1988). Unfortunately, the sample sizes are very small.

Thirteen elements were determined by electron microprobe analysis, to wit, Ca, Ba, Cl, S, Sr, Fe, Mn, Na, Mg, Al, Si, P, and O. Thirty analyses were made on each carapace at points situated according to a predetermined pattern. The results for each individual (being the average of 30 values for each element) were expressed as "element atomic percent." This is a vital piece of information which you need to keep in mind for what now follows.

The main findings reported by Bodergat et al. (1993) were:

(a) There are no significant differences between chemical compositions of carapaces from the different stations.

(b) There are strongly manifested seasonal differences due, it was surmised, to the facilitated incorporation of Mg during the winter months.

(c) The chemical composition of summer individuals was interpreted as being due to the influence of salinity fluctuations on the shell and to the supply of terrigenous sediment.

The data were analysed by multivariate statistical methods, namely, discriminant functions and "normalized principal component analysis". The latter procedure does not seem to be principal component analysis, as is usually conceived, but rather, some kind of Q–R-mode application of latent roots and vectors. A quick appraisal of the results shows, moreover, that an inappropriate model for covariances was used with the consequence that the interpretations arrived at by Bodergat et al. (1993) are not unchallengeable. A full analysis of these data has appeared in the journal *Environmetrics* (Reyment, 1996).

The Data

The data are expressed as percentages, but if you examine, say, the data in the file *leptocva.dat*, you will observe that although the rows, each with 12 entries, have a constant sum, this is not 100. You can test each of the three samples with the program *propmat.exe*. The reason why the data to not sum to the expected 100% is that they have been "doctored". There are 12 entries but more elements were actually determined. The missing entries are for oxygen and, possibly, also carbon.

Now that we have established the fact that the data are compositional, the next step is to establish rules for an appropriate multivariate model. This is where the analysis published by Bodergat et al. (1993) goes awry insofar as the vital consideration of "closure" does not seem to have been taken into account. Eliminating variables from a compositional data-set does not help things, since the parts constituting such data are related in a complicated manner that does not exist for "free" variables.

Principal component analyses

One of the problems bothering the original analysts was the difference in variability from variable to variable. The following computations were therefore made on the log-ratio correlation matrix, which has the effect of "stabilizing" the variation.

We consider first the constrained model for the Spring data. There are 13 observational vectors and 12 parts.

The analysis for the log-ratio transformed data

latent roots

5.4706 1.9393 1.6030 1.1547 0.7117 0.5702 0.2946 0.1610 0.0747
0.0149 0.0053

latent vectors

	1	2	3	4	5	6	7	8	9	10	11	12
1	0.40614	−0.06614	0.10312	0.09289	−0.17667	0.11051	0.00725	0.41703	−0.12384	0.29717	0.67922	0.17040
2	0.38510	−0.16283	−0.00639	−0.00982	0.29110	0.07343	−0.27674	−0.30185	−0.69693	−0.09298	−0.13656	0.23284
3	−0.22007	0.08412	−0.55893	0.26318	0.23596	0.21639	−0.44218	0.29850	0.15282	−0.18952	0.06291	0.33537
4	0.19997	0.56158	−0.13943	0.23220	−0.10786	0.17762	0.00823	−0.52712	0.18048	0.43034	−0.00936	0.17979
5	0.32422	−0.36225	0.08414	−0.09055	−0.36666	0.02361	−0.36844	0.13378	0.34175	0.22301	−0.49645	0.21621
6	0.08124	−0.03148	−0.56010	−0.50552	0.08327	−0.50053	0.22699	0.01740	−0.01625	0.25464	0.02306	0.21958
7	−0.05869	0.25699	0.47629	−0.49807	0.50412	0.13808	−0.13810	0.09651	0.23162	0.08195	0.03698	0.30374
8	−0.06288	−0.59251	0.01179	0.30488	0.41028	0.13911	0.39264	−0.22342	0.25933	0.20855	0.05555	0.21515
9	0.38964	0.04876	0.10544	0.13777	−0.00217	−0.42373	−0.00741	−0.21144	0.36176	−0.60701	0.21234	0.21673
10	−0.28033	0.16304	0.29867	0.42574	−0.01983	−0.51617	0.05246	0.22857	−0.23263	0.19445	−0.21177	0.40717
11	−0.34947	−0.09696	0.03804	−0.24632	−0.49880	0.27969	0.18502	−0.21151	−0.13405	−0.24220	0.12220	0.55568
12	0.35591	0.24285	−0.08903	0.04912	0.06075	0.30051	0.57465	0.38587	−0.04600	−0.22852	−0.39955	0.13006

Bodergat et al. (1993) attempted to ordinate the 41 specimens by the principal component scores of the three samples pooled. This is inappropriate for the problem at hand given that in addition to the difficulty implied by the constraint, the data are not homogeneous with respect to variances and covariances; hence, any form of "communal principal component analysis" (i.e. principal components on the pooled within-groups matrices) is unsuitable.

Log-contrast canonical variate analysis and generalized distances

The recommended procedure is to use Aitchison's (1986) log-ratio transformation of the data matrix by means of the program *cvaconst.exe*. You can run the two sets of computations now, firstly for the data grouped according to the seasons at which they were sampled and, secondly, according to the sites. The two data sets are in files

leptall.dat for the seasons, and,
leptostn.dat for the three sampling sites.

The computations are made on the covariances.

Comments on the analyses

You should first obtain the full set of results for both seasons and stations (sites) after which perusal of the following notes will probably prove useful.

  1 Look at the generalized statistical distances. These all yield significant values of T^2, thus indicating that there are genuine differences between samples; that is, for both seasons as well as sites. This is formally confirmed by the results for the multivariate analysis of variance (listed in the output as "test of equality of means").

  2 The plots of the canonical variate scores (note, that there are only two canonical vectors, since there are only three groups) show excellent group-ordination. In the case of the seasons, there is no overlap between groups, whereas for the sites, there is slight overlap between the North Sea and the English Channel. Note, that this aspect of canonical variate analysis is, in effect, a multiple discriminant function analysis.

  3 The pooled principal component analyses for the correlations and covariances and for both groupings of the data, do not yield an informative result. The points plotted for the different categories merge over the whole field of the graph.

  4 One of the findings made by the original analysts was that the chemical composition of the shell is influenced by salinity (Na, Cl), terrigenous components (Si, Al, Fe) and the metabolic role of Mg. This may well be true, but the segregation into functional components that should be apparent in the appropriately formulated chemometric approach does not appear in the clear-cut manner claimed by Bodergat et al. (1993). You can check this yourselves by examining the first principal components provided as an appendage to the canonical variate analysis. Furthermore, if you test the vectors for collinearity, you will find that the angles formed between vectors are relatively small, thus illustrating that the first latent vectors of the correlation matrices are roughly collinear. Use the program **compvec.exe** and the data file **leptovec.dat**. This program asks you to type in the name of the file in a box that comes up on the screen. When you have done this, it will ask for a name to be given to a file to hold the results. The data-file consist of pairings of the first latent vectors of the relevant correlation matrices. The smallest angle of 9.36° is for the comparison between summer and winter, but this is not due to close agreement in Mg-loadings alone, since that for Spring is about the same. We have not concerned ourselves with the lesser latent vectors, not only on account of the small samples, but also because of the instability in coefficients usually associated with such non-Gaussian data.

Conclusions

We are now in a position to examine the validity of the conclusions arrived at by the original investigators. The element Ca was excluded from the original work owing to its "dominance" and, presumably, because it was felt that eventual environmental differences would be more specifically addressed by the minor elements. It was retained for present analytical purposes. You can test what happens if Ca is eliminated from the analyses. We found that the results were essentially the same with respect to the question of the ordinations, although these were less sharply manifested.

1 *Each season sets its stamp upon the shell-chemistry of the species.* This is borne out by the present analysis, and even more persuasively than in the original study.
2 *There are no significant differences between the samples taken from the three localities.* This is false. The canonical variate analysis shows quite clearly that the three samples are indeed different. This is supported by the MANOVA and by the generalized statistical distances, all of which transform to highly significant values of T^2.

We have underscored the importance of selecting an apt statistical model for analysing compositional data, and this has been indeed one of the main themes pervading this tract. What happens in the case of canonical variate analysis if one just stumbles ahead by way of the inappropriate procedure. You can test this yourselves by running the untransformed data in **cva.exe**. You will find that as before, the ordination for sites gives excellent separation, in fact, slightly better than before. Moreover, the generalized statistical distances are highly significantly different. The separation for seasons is also excellent. This brings out an important practical point, namely, the robustness of the ordination aspect of the method of canonical variates. No matter whether you use the right procedure or the "wrong" one, the relationships between individual points will not be much altered by an analysis concerned with but a few groups. If you are also interested in the indications provided by the canonical vectors, then it is important to select the appropriate procedure. In the present example, the original authors used a low-grade ordinating method (and an inappropriate statistical model) which accounts for their failure to identify fully the differentiation so well manifested in their material.

There are many more points of analytical interest, but we leave their elucidation to your newly acquired skills. The case history briefly reviewed here is an example of the rapidly growing field of *geological environmetrics*.

MULTIPLE GROUP PRINCIPAL COMPONENT ANALYSIS

Introduction

One of the newer methods of multivariate analysis to emerge is known as *Common Principal Component Analysis* (Flury, 1984, 1988, 1995), a procedure for the simultaneous principal component analysis (PCA) of several groups. Although common principal components can be computed both for covariance and correlation matrices, the computations for large-sample standard errors of the latent roots and vectors apply only in the former case.

Standard PCA is normally a one-sample method, apart from one special case shortly to be mentioned. However, a variety of ad hoc applications to data-analysis in the geological and biological sciences abound in which this fact is not understood or just conveniently glossed over. The following account has been made fairly comprehensive, owing to the fact that the method is new and as yet cannot be consulted in commonly available handbooks.

Basis of common principal component analysis

The method of common principal components can be introduced in relation to similarities in covariance matrices.

1 Firstly, there is the simplest case when all covariance matrices are equal.
2 The second degree in this hierarchy is when the covariance matrices are involved in a proportionate relationship; i.e.

$$\Sigma_i = a_i \Sigma_i, \quad i = 2, \ldots, k \tag{5:6}$$

for some positive constant α_i for k groups.
3 The third category encompasses the CPC model, now to be introduced.

Although the method of CPCA is quite general, it is best introduced in terms of the two-group situation. Let there be two samples of an object on which multivariate observations have been made. The question asked is whether a unique common transformation for both groups can be estimated? The reason for wanting to try such an analysis is that slight differences in covariance matrices could well be due to sampling variation and, therefore, lack intrinsic analytical significance. A standard principal component analysis, with comparisons of latent vectors would then not be really appropriate. The model for two samples can be expressed simply as

$$\Sigma_1 = \beta\Lambda_1\beta^T$$
$$\Sigma_2 = \beta\Lambda_2\beta^T \qquad (5:7)$$

where the λ_i are diagonal and the Σ_i are positive definite.

The relationships in eqn. (5:7) are, in effect, saying that the two dispersion ellipsoids are identically oriented, but differently inflated. Geometric realizations of this, and related situations, are given in Reyment (1969, Fig. 1). This is a special case of heterogeneity in covariance matrices, the general case being that not only are the dispersion ellipsoids differently inflated but their major axes are not parallel. It is important to keep in mind that this latter situation is not germane to the common principal component model. The interpretational relevance of the CPC representation is that there may be some interesting proportional relationship between covariance matrices, such as can occur in biological work.

By a theorem for the simultaneous decomposition of two positive definite symmetric matrices (Bellman, 1960) there exists a non-singular $p \times p$ matrix **B** such that:

$$\mathbf{B}^T\Sigma_1\mathbf{B} = \mathbf{I}_p$$

and

$$\mathbf{B}^T\Sigma_2\mathbf{B} = \Lambda \qquad (5:8)$$

where Λ is a diagonal matrix. Hence

$$\Sigma_1^{-1}\Sigma_2\mathbf{B} = \mathbf{B}\Lambda \qquad (5:9)$$

That is, the columns of **B** are the latent vectors of the matrix product $\Sigma_1^{-1}\Sigma_2$ and the diagonal of Λ contains the corresponding latent roots. (N.B. eqn. (5:9) exposes why compositional data will not fit the *normal* CPC model with its requirement of positive definite matrices.) The statistical implication of the simultaneous decomposition theorem is that it provides a convenient vehicle for obtaining uncorrelated variables in two populations. Thus if

$$\mathbf{U} = \mathbf{B}^T\mathbf{X} \qquad (5:10)$$

then the covariance matrix of **U** is $\mathbf{I}_p$ in the first population and Λ in the second population. This is *almost* but not quite **exactly** a generalization of principal component analysis to two groups. The catch is that **B** is, in general, not exactly orthonormal and hence the definition of the principal component transformation as a rotation of the coordinate system does not invariably apply to the multiple group situation as expressed in the CPC formulation.

There are some special situations for two groups in which the latent vectors of $\Sigma_1^{-1}\Sigma_2$ are orthogonal. For example, the simple circumstance where the latent vectors of both covariance matrices are identical, the latent vectors of $\Sigma_1^{-1}\Sigma_2$ are mutually orthogonal, where the product $\Sigma_1^{-1}\Sigma_2$ is symmetric, and where $\Sigma_1\Sigma_2 = \Sigma_2\Sigma_1$.

The solution for two groups extends simply to several groups. The hypothesis of common principal components for p variables and k positive definite covariance matrices (each based on n_i observations) is defined as:

$$H_{CPC}: \quad \Sigma_i = \mathbf{B}\Lambda_i\mathbf{B}^T \quad (i = 1, \ldots, k) \tag{5:11}$$

$\mathbf{B}$ is an orthonormal matrix and Λ_i is diagonal. Hence, the CPC transformation can be interpreted as being a rotation that produces new variables that are **as uncorrelated as possible**. The special circumstance referred to above in which a standard multiple group PCA, using the pooled within-group covariances, is appropriate occurs when all covariance matrices are equal.

Assessing the adequacy of the cpc model

Using the maximum likelihood estimates of the $\mathbf{B}$, the sample CPCs may be defined as

$$\mathbf{U} = \hat{\mathbf{B}}^T\mathbf{X} \tag{5:12}$$

The *sample covariance matrix* of $\mathbf{U}$ in group i (eqn. (5:12)) is

$$\mathbf{F}_i = \hat{\mathbf{B}}^T\mathbf{S}_i\hat{\mathbf{B}}_i \tag{5:13}$$

and

$$\Lambda_i = \text{diag } \mathbf{F}_i \tag{5:14}$$

holds.

An effective way of judging the suitability of the common principal component model is to examine the corresponding sample correlation matrices obtained from (5:13).

$$R_i = \hat{\Lambda}^{-1/2}\mathbf{F}_i\hat{\Lambda}^{-1/2} \quad (i = 1, \ldots, k). \tag{5:15}$$

The off-diagonal elements of these correlation matrices (5:15) are expected to be close to nought, that is, the matrices should approximate the identity matrix $\mathbf{I}_p$ under the hypothesis H_{CPC}. Marked (significant) deviations from zero correlation indicate that the CCP model is likely to be inappropriate.

Flury (1988) also indicated how to compute large-sample standard errors for the latent roots and latent vectors of common principal component analysis, based on results of Anderson (1963). The standard error for latent roots is given as (cf. Anderson, 1963, 1984):

$$s(\hat{\lambda}_{ij}) = (2/n_i)^{1/2} \hat{\lambda}_{ij} \qquad (5:16)$$

where the n_i denote individual sample sizes. There will, therefore, be a set of standard errors for the latent roots computed for each sample. On the other hand, there is only one set of standard errors for the principal components **B**, as indicated by eqn. (5:13), since these are common to all k groups. If desired, confidence intervals can be easily obtained for the latent roots (Anderson, 1963).

The large-sample standard errors for the common principal component coefficients are yielded by the following equations numbered (5:17) and (5:18).

$$s(\hat{\beta}_{mh}) = \left(\frac{1}{n} \sum_{i=1}^{p} \theta_{jh} \hat{\beta}_{mj}^2 \right)^{1/2} \qquad (j \neq h) \qquad (5:17)$$

where θ_{jh} is the harmonic mean of

$$\hat{\theta}_{jh}^{(i)} = \left(\frac{n}{n_i} \frac{(\hat{\lambda}_{ij} \hat{\lambda}_{ih})}{(\hat{\lambda}_{ij} - \hat{\lambda}_{ih})^2} \right) \qquad (j \neq h) \qquad (5:18)$$

Equation (5:17) is the sum over p variables of the product of the harmonic mean of the product of latent roots divided by their differences and the square of the common principal components.

The log-likelihood ratio statistic for testing H_{CPC} is the well known test for homogeneity of multivariate Gaussian populations (see, for example, Seber, 1984), to wit,

$$\sum_{i=1}^{k} n_i \log(\det \hat{\Sigma}_i) - n \log(\det \mathbf{S})) \qquad (5:19)$$

which is a large-sample chi-square on $(k - 1)p$ degrees of freedom. Here, the Σ_i are obtained from eqn. (5:11) and matrix **S** is the pooled sample covariance matrix defined as in the following equation:

$$\mathbf{S} = n^{-1} \sum_{i=1}^{k} n_i \mathbf{S}_i$$

The adequacy of the CPC model H_{CPC} is tested by means of the covariances $\mathbf{F}_i$ on $(k-1)p(p-1)/2$ degrees of freedom (eqn. 5:20).

$$\chi^2 = \sum_{i=1}^{k} n_i \log \frac{\det(diag\ \mathbf{F}_i)}{\det \mathbf{F}_i}. \qquad (5:20)$$

This statistic provides a measure of how well the k covariance matrices can be subjected to simultaneous diagonalization.

Common principal component analysis requires careful attention to detail. It is necessary to keep a close watch on what is happening as the analysis progresses and to make sure that the model is really appropriate to the problem and the data. **Box 17** summarizes simple applications of common principal component analysis. The program for doing the computations *cpca.exe*, is based on code generously supplied by Professor B. Flury; it is accessed as follows:

Instructions for using the program **cpca**

Line 1: Number of covariance matrices in problem (up to a maximum of 10)

Line 2: The sample-sizes corresponding to these matrices.

Line 3+ The matrices to be analysed, sequentially ordered.

Output details

    1 A standard principal component analysis for each matrix.
    2 A test for a valid common principal component situation.
    3 The matrices **F** and corresponding correlations, **R**.
    4 Standard errors for the common principal component latent roots and vectors.
    5 Homogeneity of covariance matrices test.

Obtaining the matrix input for running CPCA

The program **covpc.exe** computes the input matrices for the common principal components. The input information for using **covpc** is

Line 1: 1 the number of samples
        2 the number of variables
        3 type 1 for logarithms of the data otherwise type 0

Line 2: the sample size (this is to be supplied in front of each matrix) and the data-matrix.

The relevant output from **covpc** for insertion into **cpca** is stored in file *cpcinput* in a form that permits immediate insertion into the latter program. There is a set of trial-data in the file *gerris.dat*, consisting of morphometric measurements on species of gerrids (water-striders) and taken from Klingenberg (1996).

The data used in the following illustration for interpreting common principal components are available in the file *cyprid.dat*, being three measures on the carapace of the living freshwater ostracod species *Cypridopsis vidua* (Müller).

Box 17: An example of common principal component analysis

Program: **cpca**

Data: **cyprid.dat**

The best way to understand what CPCA tries to do is by means of a worked example. The main steps are introduced in the form of an environmetric problem, extracted from Reyment and Brännström (1962).

Environmental effects on Cypridopsis vidua *(Müller)*

This example demonstrates a case in which the CPC model is appropriate. The data consist of log-transformed measurements on the length (L), height (H) and breadth (B) of the carapace of the freshwater ostracod species, *Cypridopsis vidua* (Müller), taken on samples of individuals reared in three different environments, to wit, an environment with normal freshwater conditions ($N = 36$), an environment enriched in calcium carbonate ($N = 53$), and a stagnant environment ($N = 30$). One of the aims of the study undertaken by Reyment and Brännström (1962) was to see in what manner the morphology of the ostracod carapace displays evidence of reaction to environmental conditions. The covariance matrices and the corresponding usual principal component results for the three samples are listed consecutively below:

1 Normal environment

$$\begin{bmatrix} 0.1234 & 0.0778 & 0.1117 \\ 0.0778 & 0.1235 & 0.0929 \\ 0.1117 & 0.0929 & 0.1964 \end{bmatrix}$$

Standard latent roots

0.34381   0.060212   0.03927

Standard latent vectors

	1	2	3
L	0.52545	−0.04548	0.84960
H	0.48138	−0.80747	−0.34095
B	0.70154	0.58814	−0.40239

2 Calcareous environment

$$\begin{bmatrix} 0.9244 & 0.4640 & 0.8419 \\ 0.4640 & 0.4066 & 0.5395 \\ 0.8419 & 0.5395 & 0.8677 \end{bmatrix}$$

Standard latent roots

2.04523   0.14108   0.01238

Standard latent vectors

	1	2	3
L	0.65061	−0.63279	−0.41987
H	0.39730	0.75481	−0.52193
B	0.64719	0.17275	0.74249

3 Stagnant environment

$$\begin{bmatrix} 0.1534 & 0.0705 & 0.0630 \\ 0.0705 & 0.1001 & 0.0740 \\ 0.0630 & 0.0740 & 0.1365 \end{bmatrix}$$

Standard latent roots

0.26974   0.08271   0.03754

Standard latent vectors

	1	2	3
L	0.62801	−0.73311	−0.26103
H	0.51533	0.14044	0.84540
B	0.58312	0.66544	−0.46600

The first latent vectors for the three-covariance matrices do not differ greatly from each other, whereas the second and third vectors are quite different. This indicates that a CPC model could only be expected to be reasonable for the first principal component. This could be interpreted as fitting the CPC "one-vector" allometric model of Klingenberg (1996) for multiple groups (see also Hopkins, 1966).

The log-likelihood test for a valid CPC model yields a value of $\chi^2 = 10.52$ for 6 degrees of freedom. For comparison, the 95% quantile of chi-square on 6 degrees of freedom is 12.59, which indicates a reasonable fit of the model (the value does not attain statistical significance). Comparing the CPC model with that for equality of all covariance matrices, the value of the statistic (5:19) is a highly significant chi-square of 95.588 for 9 degrees of freedom, a result that provides additional backing for the CPC model (the 95% quantile is 16.92). The explanation of this result can be seen from inspection of the ordinary principal components and corresponding variances which indicates that the differences between dispersion ellipsoids lie with differing inflations along the principal diagonal (cf. Reyment, 1969, Fig. 1) for the normal and calcareous environments and that the order of the latent vectors for the stagnant environment is reversed such that vector 3 corresponds to the second vector of the other two samples. A reasonable explanation of this is the occurrence of sphericity in these two roots (thus leading to a fortuitous reversal in the order of the latent vectors).

Notwithstanding that the tests for H_{CPC} do not lead to the rejection of a CPC model, it is instructive to examine the correlations defined in equation (5:15) between estimated CPCs. These are listed below:

1 Normal environment

$$\begin{bmatrix} 1.0000 & 0.2629 & 0.1086 \\ 0.2629 & 1.0000 & -0.1380 \\ 0.1086 & -0.1380 & 1.0000 \end{bmatrix}$$

2 Calcareous environment

$$\begin{bmatrix} 1.0000 & -0.1444 & -0.0003 \\ -0.1444 & 1.0000 & -0.0001 \\ -0.0003 & -0.0001 & 1.0000 \end{bmatrix}$$

3 Stagnant environment

$$\begin{bmatrix} 1.0000 & 0.1035 & -0.1667 \\ 0.1035 & 1.0000 & 0.3465 \\ -0.1667 & 0.3465 & 1.0000 \end{bmatrix}$$

The only relatively high correlation, 0.3465, occurs for the stagnant environment (0.26 is not high from a practical viewpoint). It could be argued that the higher values could derive from atypical observations, but such is not the case, for the distributions are multivariate Gaussian (Reyment and Brännström, 1962).

The common principal component vectors were computed to be

	1	2	3
L	0.6239	−0.6591	−0.4198
H	0.4280	0.7377	0.5219
B	0.6537	0.1459	0.7424

and the CPC latent roots, sample by sample, are

1	0.33931	0.05612	0.04786
2	2.04200	0.14432	0.01238
3	0.26689	0.06382	0.05928

The latent values are quite close to the ordinary principal component counterparts, which is a further indicator of the usefulness of the CPC model for these data.

The standard errors of the common principal component coefficients (by eqns (5:17) and (5:18)) are

	1	2	3
L	0.01584	0.03586	0.01960
H	0.01933	0.03552	0.02339
B	0.02519	0.02305	0.03336

These values are small and of the same order of magnitude, thus suggesting that the principal component coefficients are probably relatively stable.

The standard errors for each set of CPC latent roots by eqn. (5:16) are

	1	2	3
Normal environment	0.07998	0.03230	0.01128
Calcareous environment	0.39667	0.00240	0.02804
Stagnant environment	0.06891	0.01648	0.01531

A final exercise that is usually of interest concerns the reconstitution of the input covariance matrices and examining residuals, since this provides additional information on how well the CPC model can be applied to the data. The results of the relevant calculations are:

1 Normal environment

	Reconstructions			Residuals		
	L	H	B	L	H	B
L	0.1627	0.0796	0.1163	−0.0393	−0.0018	−0.0046
H	0.0796	0.1035	0.0783		0.0200	0.0146
B	0.1163	0.0783	0.1769			0.0195

2 Calcareous environment

	L	H	B	L	H	B
L	0.8598	0.4779	0.8152	0.0646	−0.0139	0.0267
H	0.4779	0.4561	0.5822		−0.0495	−0.0427
B	0.8152	0.5822	0.8827			−0.0150

3 Stagnant environment

	L	H	B	L	H	B
L	0.1409	0.0564	0.0832	0.0125	0.0141	−0.0202
H	0.0564	0.0985	0.0563		0.0016	0.0177
B	0.0832	0.0563	0.1505			−0.0140

Agreement with the input covariance matrices is reasonably satisfactory, the most divergent sample being that for the calcareous environment.

COMMON PRINCIPAL COMPONENT ANALYSIS FOR COMPOSITIONAL DATA

It has already been pointed out that the usual form of the CPCA model (Flury, 1988) will not support compositional data (i.e. multivariate observations, each vector of which has a constant and identical sum) owing to the singularity of the covariance matrices and the obstacle inherent in eqn. (5:9). A suitable adaptation of CPC can, however, be constructed using the log-ratio covariance matrix, the elements of which are, defined as (cf. p. 24) and Reyment (1997):

$$\Sigma = [\text{cov}\{\log(x_i/x_D), \log(x_j/x_D)\}]; \quad i, j = 1, \ldots, d$$

where D denotes the number of parts (i.e. ingredients) in the composition and $d = D - 1$; Σ is the covariance matrix of a d-dimensional random vector

$$y_i = \log(x_i/x_D) \quad (i = 1, \ldots, d)$$

The advantage of having a symmetric positive definite matrix is obtained at the cost of the loss of one of the parts, which results from the fact that the ingredients of a composition are not treated symmetrically. The centred log-ratio covariance matrix is of full dimensionality, but that matrix is singular, which constitutes a difficulty in the computation of common principal components. The log-ratio covariance matrix Σ has no restrictions on it other than the standard requirement of non-negative definiteness.

Based on the results of Aitchison (1986, p. 192), a log-contrast CPCA can be constructed using k log-ratio covariance matrices as in the usual situation as outlined in the foregoing.

Instructions for using the program **cpcconst**

Program *cpcconst.exe* computes a constrained common principal component analysis for up to 10 samples simultaneously and 20 variables. It uses the log-ratio covariance matrix. The matrices, in suitable format, are obtained from **covpc** (p. 194).

Line 1: Number of samples and number of parts on the same line.

Line 2: Title of the job.

Line 3: Number of rows in the data matrix

Line 4: The data-matrix of compositions

Steps 2 to 4 are repeated for each sample.

The training set is in file *aitchcp.dat*, which comprises the two fictitious data-sets of *hongite* and *kongite* (Professor Aitchison was for many years Professor of Mathematical Statistics at the University of Hong Kong, hence the choice of names).

A COMPOSITIONAL EXAMPLE

The following briefly expounded example is based on constructed data, obtained from Aitchison (1986). The observations consist of five fictitious mineral species determined on 25 samples each of Boxite and Coxite (a whimsical allusion to a classical paper of mathematical statistics by G. E. P. Box and D. R. Cox). An initial test of equality of covariance matrices gave a highly significant value.

A compositional example

The "standard" principal component analysis for Boxite yields the following latent roots and vectors (i.e., we have extracted the latent roots and vectors of the respective matrices without recourse to the common principal component model).

Standard latent vectors

	1	2	3	4
1	−0.2408	0.4028	−0.5572	0.6849
2	−0.6395	0.2552	0.6988	0.1935
3	0.7252	0.4462	0.4103	0.3263
4	−0.0836	0.7572	−0.1809	−0.6219

Standard latent roots

1.44589   0.30508   0.01014   0.00429

The corresponding results for Coxite are:

Standard latent vectors

	1	2	3	4
1	−0.1924	0.3621	0.8092	0.4205
2	−0.6024	0.3796	−0.5438	0.4440
3	0.7745	0.3931	0.2219	0.4429
4	−0.0080	0.7550	0.0008	−0.6555

latent roots

1.66174   0.34193   0.01022   0.00014

The two sets of results are quite close with respect to the elements of the latent vectors, whereas the corresponding first two latent roots for Coxite are greater. There is therefore a reasonable indication that a common principal component model is worth considering. The pro forma correlation matrices obtained under the CPC assumption are:

Boxite

1	1.0000	−0.3037	−0.0198	0.2414
2	−0.3037	1.0000	−0.1506	−0.0944
3	−0.0198	−0.1506	1.0000	−0.0946
4	0.2414	−0.0944	−0.0946	1.0000

```
              Coxite
1      1.0000   0.0226  0.0194  −0.2387
2      0.0226   1.0000  0.0230   0.0144
3      0.0194   0.0230  1.0000   0.0928
4     −0.2387   0.0144  0.0928   1.0000
```

The data for Coxite clearly fit the model better than do those for Boxite; the largest pro-forma correlation is −0.3, which is not a very serious deviation.

The test for equality of covariance matrices yields a chi-square of 54.3 which, with 16 degrees of freedom, is highly significant; this result, therefore, does not reject the CPC model. The log-likelihood test for a valid CPC model gives a chi-square of 6.3697, which for 6 degrees of freedom, is not significant and hence indicates agreement with the CPC model.

The common principal component coefficients (i.e. latent vector elements), listed by columns, are

```
         1         2         3         4
1    −0.2143   0.3868    0.7908    0.4231
2    −0.6199   0.3223   −0.5622    0.4422
3     0.7532   0.4233   −0.2397    0.4425
4    −0.0480   0.7531   −0.0307   −0.6553
```

and the corresponding latent roots are:

```
Boxite   1.44227   0.30792   0.00992   0.00530
Coxite   1.65799   0.34502   0.01088   0.00014
```

The standard errors for the common principal component coefficients are as follows:

```
        1         2         3         4
1   0.03314   0.02719   0.01430   0.01842
2   0.02733   0.05309   0.01511   0.01316
3   0.03494   0.06230   0.01745   0.00598
4   0.06196   0.00482   0.02485   0.00312
```

Standard errors for latent roots; Boxite

0.40794   0.08709   0.00281   0.00150

Standard errors for latent roots; Coxite

0.46895   0.09759   0.00308   0.00040

The matrix of residuals obtained by subtracting the reconstituted covariance matrix from the input matrix is for Boxite:

	1	2	3	4
1	0.0191	0.0241	−0.0158	0.0185
2	0.0241	0.0259	−0.0035	0.0182
3	−0.0157	−0.0035	−0.0518	−0.0308
4	0.0185	0.0182	−0.0308	0.0067

That for Coxite is:

	1	2	3	4
1	−0.0214	−0.0232	0.0123	−0.0212
2	−0.0232	−0.0211	0.0025	−0.0272
3	0.0123	0.0025	0.0470	0.0410
4	−0.0212	−0.0272	0.0410	−0.0044

The elements of these two matrices are relatively small and, on the whole, quite similar. In conclusion, it may be suggested that a common principal component analysis is justified and that the result indicates a simplification to the first two principal components without loss of important descriptive variability in the material.

CONCLUDING REMARKS

The method of common principal component analysis provides an interesting alternative in the analysis of some types of problems in geobiology. The use of CPCA requires close attention to statistical detail and it is not likely that it will prove applicable to a great number of situations in the Earth Sciences. At present, the main geological sphere of interest for CPCA lies with palaeontology, and its practical applications, palaeoecology and evolutionary studies. An important field is that of allometry on the Hopkins (1966) model (Klingenberg, 1996). An advantage of the CPC model is that it provides an attractive possibility for comparing and contrasting graphs of multiple group scores, as is well illustrated by Klingenberg in the above-cited publication.

In summary, the main points to be kept in mind on examining a likely candidate data-set for CPCA are:

1 Test the covariance matrices for equality. If the covariances are found to be equal, then a multiple group principal component analysis (i.e. a standard principal component analysis) can be made on the pooled covariance matrices. In this case the CPC model is obviously not appropriate.

2 If the CPC correlations tend to be rather large, this could be due to the presence of atypical and influential observations in the original data set (Krzanowski, 1987a, 1987b; Reyment, 1991). Usually (but not invariably), graphical scanning of the data will uncover such observations.

3 It is useful to reconstitute the covariance matrices under the CPC model and compare these with the original input matrices. Large residuals indicate that the model does not fit well. The third example for compositional data yields small residuals and is therefore a good fit to the model.

4 Be on the watch for the presence of multiple roots, that is, simultaneous sphericity in two or more principal components. If identical roots occur, the connected latent vectors will be poorly defined.

The question of assessing stability in CPC latent vectors has still to be probed. The procedures available for robust estimation of principal components (Campbell, 1980, 1984; Seber, 1984) are not yet directly applicable, nor are shrunken estimators as used in robust canonical variate analysis (Campbell and Reyment, 1978); there is clearly an opening for further developments here. In the meantime, the wisest course to take is to restrict CPC to multivariate Gaussian data sets. Atypical and influential observations should be isolated from each sample in turn by graphical means and by some such technique as Krzanowski's (1987a, 1987b) cross-validation.

The examples presented are realistic case studies which illustrate the kinds of problems likely to confront the analyst.

STABILITY IN CANONICAL VECTOR ELEMENTS

In canonical variate analysis (the two-group specification of which is the familiar discriminant function) ascertaining the efficiency of variables in discriminating between groups can be a problem. In discriminant analysis, it can be shown that high correlation within groups, combined with between-group correlation of the opposite sign, leads to greater group separation and a more powerful test than when the within-groups level of correlation is low. There is a strong formal connection between discriminant analysis and multiple regression and, as suggested by Campbell (1980), the instability associated with highly correlated regressor variables carries over to canonical variate analysis. This implies that any interpretations based on the relative magnitudes of the elements of canonical vectors can be misleading. In short, when the sum of squares between the means along a particular latent vector of the within-groups dispersion matrix is small, and the corresponding latent root is also small, instability of the coefficients will probably result.

Notwithstanding that the reification of canonical elements has been largely discredited, it is still of interest to compare such vectors in connexion with the regional validity of analytical findings.

Description of the method of shrunken canonical estimators

The within-groups sums of squares and cross products matrix $\mathbf{W}$ on n_w degrees of freedom, and the between-groups sums of squares and cross products matrix $\mathbf{B}$, are computed in the usual manner of canonical variate analysis. Although not necessary for the application of the method, it is usually advisable to standardize $\mathbf{W}$ to correlational form (whereby dispersion hyperellipsoids become hyperspheres), with the same scaling for $\mathbf{B}$.

$$\mathbf{W}^* = \mathbf{S}^{-1}\mathbf{W}\mathbf{S}^{-1}$$
$$\mathbf{B}^* = \mathbf{S}^{-1}\mathbf{B}\mathbf{S}^{-1} \qquad (5:21)$$

The latent roots $\mathbf{E}$ and vectors $\mathbf{U}$ of $\mathbf{W}^*$ are then computed

$$\mathbf{W}^* = \mathbf{U}\mathbf{E}\mathbf{U}^T \qquad (5:22)$$

The latent vectors are then scaled by dividing them by the square root of their corresponding latent roots. Shrunken estimators are formed by adding shrinkage constants k_i to the latent roots before carrying out the scaling to produce matrix $\mathbf{U}^*$.

$$\mathbf{U}^* = \mathbf{U}(\mathbf{E} + \mathbf{K})^{-1/2} \qquad (5:23)$$

The next step is to construct the between groups matrix $\mathbf{G}$ in the space of the within-groups principal components:

$$\mathbf{G} = \mathbf{U}^{*T}\mathbf{B}^*\mathbf{U}^* \qquad (5:24)$$

The i-th diagonal element of $\mathbf{G}$ is the between-groups sum of squares for the i-th principal component. If no shrinkage constants have been inserted, the latent roots and vectors of $\mathbf{G}$ are the usual canonical roots $\mathbf{f}$ and vectors, $\mathbf{A}$. Where shrinkage constants have been inserted (5:23), shrunken (or generalized ridge estimators) $\mathbf{C}$ are yielded by:

$$\mathbf{C} = \mathbf{U}^*\mathbf{a} \qquad (5:25)$$

With respect to the selection of shrinkage constants, an "infinitely large value" confines the canonical solution to the subspace orthogonal to the vector, or vectors, affected by the addition. As a rule of thumb, one may say that when the between-groups sum of squares is small (say, less than 5% of the total variation between groups), and the corresponding latent root is also small (say, less than 1–2% of the trace of $\mathbf{W}$), then shrinking of the principal component will probably be of analytical value.

Campbell and Reyment (1978) gave a worked example for the shrinkage treatment of a Late Cretaceous foraminifer, *Afrobolivina afra*, already introduced on p. 116. The program is called **shrinkcv.exe**, the operating details for which now follow:

Instructions for using the program **shrinkcv**

Line 1: 1 number of dimensions
        2 number of groups
        3 number of shrinkage constant vectors supplied
        4 total number of observations summed for all samples.

Line 2: the within-groups matrix in correlational form, as specified in eqn. (5:21)

Line 3: the between groups matrix in the space of the within-groups correlations, as specified in eqn. (5:21)

Line 4: the shrinkage vectors

Putting all entries to zero will give the same result as a straight canonical variate analysis. Putting an entry equal to 10 is sufficient to approximate "infinity" for the purposes of the program and hence the suppression of that direction.

Exemplification of the method

Practical experience shows that situations likely to profit from shrinking are rather uncommon. In geological work, they occur occasionally in palaeontology and, most particularly, in the study of arthropods and molluscs, two groups in which a high level of morphological integration can exist. The example we offer below is for *Veenia rotunda*, a mid-Cretaceous ostracod species from the Atlas Mountains, Morocco (cf. Reyment, 1978a). The data comprise five measurements on the carapace of the shell, length, height, distance of the adductor tubercule from the anterior margin, the length of the dorsal margin and the length of the posterior margin. The results of the analysis are presented in **Box 18**. The data are stored in file *verotshr.dat*.

Box 18: Application of Campbell's shrinkage technique to a Cretaceous ostracod species *Veenia rotunda* (Morocco).

Program: **Shrinkcv**

Data: **verotshr.dat**

Trial Number = 1

latent roots of W*

3.70278   0.79270   0.28300   0.14867   0.07284

latent vectors of W*

	1	2	3	4	5
1	0.4823	−0.2729	0.3598	0.1563	−0.7342
2	0.4519	0.2544	−0.7965	0.2837	−0.1276
3	0.3354	0.8136	0.4470	0.0585	0.1495
4	0.4621	−0.4301	0.1700	0.4123	0.6345
5	0.4870	−0.1180	−0.0865	−0.8495	0.1406

The "smallest" latent root contains two large elements (variables 1 and 4) with opposite signs. This makes this 'direction' a possible candidate for elimination with the likelihood that stability in the canonical vectors will result.

The input shrinkage constants

0.00   0.00   0.00   0.00   0.00

This line indicates that none of the between-groups directions is to be suppressed; i.e. a standard canonical variate computation will be made.

Between groups sums of squares for principal components

0.45415   0.03330   0.75174   0.39589   0.17378

The smallest between-groups sum of squares is the second entry. The next smallest is the fifth entry. However, this last value is greater than what usually causes stability problems.

Usual canonical variate results

1	2	3	4
0.0552	−2.0511	−1.3387	−0.1174
0.9082	0.5896	−1.1792	0.7479
−0.5879	−0.3898	0.1151	−0.0495
−1.1205	1.8717	0.0521	−0.9954
1.1430	−0.1604	1.9476	−0.3151

Usual canonical roots

1.5422 0.1960 0.0707 0.0000

Trial Number = 2

The input shrinkage constants

0.00   10.00   0.00   0.00   0.00

This line indicates that the second between-groups direction is to be suppressed.

Generalized shrunken (= ridge) estimators

1	2	3	4
0.0688	2.2596	−1.4783	0.1125
0.8926	−0.7213	−1.1327	0.4324
−0.6537	0.0835	0.1861	−0.1698
−1.0834	−1.8523	0.1114	−0.6908
1.1551	0.2880	1.9250	−0.1862

Canonical roots for estimators

1.53423   0.17367   0.07013   0.0000

Shrinking the second principal component makes very little difference to the canonical variate analysis and it may therefore be interpreted as a redundant direction.

Trial Number = 3

The input shrinkage constants

0.00   0.00   0.00   0.00   10.00

Generalized shrunken (ridge) estimators

1	2	3	4
−0.5398	−0.0433	−0.3902	−0.4828
0.8389	1.2750	−0.3904	0.6384
−0.4959	−0.7507	−0.8038	0.2575
−0.6300	0.7259	−0.5351	−0.6051
1.2751	−1.4226	1.1144	−0.2609

C-roots for estimators

1.47696   0.11850   0.04088   0.00001

Elimination of the smallest principal component perturbs all vectors and canonical roots. Its removal is therefore not motivated.

Campbell (1982) has devoted considerable attention to other aspects of stability in canonical variate analysis, including the method of M-estimation, which consists of down-weighting deviating observations and making the computations on the new data-matrix thus constructed. The details of these rather complicated analyses lie outside the scope of the present text. In addition to the above reference, the interested reader can consult Campbell (1984), Campbell and Reyment (1980), Krzanowski (1988), Reyment (1991), and Seber (1984). If you are contemplating doing canonical variate analyses in the face of divergent data sets, we strongly recommend study of Campbell's publications.

SUMMARY

1   Canonical variate analysis.
2   Q–Q-probability plots.
3   Canonical variate analysis for compositional data.
4   A study in environmental chemistry.
5   Multiple group principal component analysis.
6   Compositional multiple group principal component analysis.
7   Stability in canonical vector elements.
8   The method of shrunken canonical variate estimators.
9   Programs and associated training sets.
    cva   gastcva.dat
    cvaconst   alkalicv.dat
    cpca   cyprid.dat
    covpc   gerrids.dat
    cpcconst   aitchcp.dat
    shrinkcv   verotshr.dat

Chapter 6

Correlating between sets

INTRODUCTION

Relatively early in the history of multivariate analysis, interest was directed towards trying to quantify the association between sets of variables having a joint distribution. Most earlier research in multivariate analytical applications was concerned with the social sciences and this initial involvement has left an indelible mark on thinking and practice, even in work done in quite different fields. Hotelling (1936) wanted to correlate between sets of different kinds of observations on the learning ability of schoolchildren. He called his procedure **canonical correlation**, pursuing in his presentation, a vague analogy with principal component interpretations. A suite of correlations is extracted from the data, analogously to the latent roots of principal components, but it is the interpretation of the results that has been the stumbling block ever since the method was introduced and even to this today, and despite numerous attempts at revamping the mode of reification of the results, canonical correlation remains something of a "maverick method" in the minds of many practitioners. We proceed now to a quick and far from comprehensive presentation of the method in order to show that in its mathematical structure, at least, canonical correlation is of the same form as canonical variate analysis of the previous section.

Just as in the case of the method of canonical variates, the problem is one of the simultaneous reduction of two symmetric matrices to diagonal form. As we saw before in the case of canonical variate analysis, the two matrices we start with are $\mathbf{T}$, the total sum of squares and deviations, and $\mathbf{W}$, the within-groups sum of squares and deviations. These matrices are symmetric positive definite. Their difference, defined as $\mathbf{B} = \mathbf{T} - \mathbf{W}$, is positive semidefinite. (This means that the determinants of $\mathbf{T}$ and $\mathbf{W}$ are positive, whereas that of $\mathbf{B}$ is nought.) We saw how the required canonical roots and vectors of canonical variate analysis are obtained from the solution of the determinantal eqn (5:4). Harris (1975, p. 141) gives a proof of the formal mathematical relationship between canonical variates and canonical correlations. Another valuable reference is that of Anderson (1984, pp. 480–502).

The mathematical structure of canonical correlation analysis is, then, based on the simultaneous reduction of two symmetric matrices to diagonal form (known technically as the Weierstrass diagonalization for the eminent German mathematician K. T. W. Weierstrass), although the input is not the same in that the sample is assumed to be from one and the same statistical population, not several, as in canonical variates. In principle, we consider two sets of variates with a joint distribution. Consider now a correlation matrix $\mathbf{R}$ (the covariance matrix can be used instead if you wish, but in that case, the variances of the variables encompassed should not be too different) consisting of correlations computed between p variables. Imagine now that these variables are of two kinds, say, q morphological dimensions in one set and r physical measures in the other. Such a matrix can be partitioned as follows:

$$\mathbf{R} = \begin{pmatrix} \mathbf{R}_{11} & \mathbf{R}_{12} \\ \mathbf{R}_{21} & \mathbf{R}_{22} \end{pmatrix}$$

All the correlations pertinent to the morphological variates are sequestered into $\mathbf{R}_{11}$, all of those relating to the physical variables into $\mathbf{R}_{22}$. The matrix $\mathbf{R}_{12} = \mathbf{R}_{21}^T$ contains the correlations between the variables of the two sets, i.e. the associations between morphological and physical traits of the sample. In this representation, $\mathbf{T}$ and $\mathbf{W}$ of canonical variate analysis are replaced respectively by $\mathbf{R}_{11}$ and $\mathbf{R}_{11} - \mathbf{R}_{12}\mathbf{R}_{22}^{-1}\mathbf{R}_{21}$.

The roots of the equation are the squares of the required canonical correlations between two new variables. Corresponding to each of these correlations there are two sets of coefficients, one matching to the one set for q variables and the other matching to the other set for r variables, that is, a new coordinate system in the space of each set of variates. These are the linear combinations of variables in each set that have maximum correlation and which are the first coordinates in the new system of coordinates. Then a second linear combination in each set is sought such that the correlation between these is the maximum of correlations between such linear combinations as are uncorrelated with the first linear associations. The process is continued until the two new coordinate systems are completely specified. Difficulties begin to arise when an attempt is made to reify these coefficients and all too often, the results seem to lack scientific sense. There is also a difficulty with respect to the canonical correlations themselves. They do not represent successively smaller portions of the total correlation between sets, as one might expect from what is achieved by the partitioning of variance in principal component analysis. In fact, it is not uncommon that a very high canonical correlation can be associated with a very minor virtual relationship between sets. Cooley and Lohnes (1971) tried to rectify

Introduction

this situation by a rather complicated, and to a certain extent perhaps, arbitrary *redundancy analysis*. Personal experience indicates that fuzzy results with confusing interpretations can be expected when:

1 The data deviate markedly from multivariate normality.
2 There are very high correlations in $\mathbf{R}_{12}$, a condition that was forecast by Campbell (1979).
3 There is great disparity in the types of variables included in one or both of the sets.
4 Variances are greatly different.

There are things that can be done in order to iron out some of these difficulties, including a suitable transformation of the data before analysis (not forgetting the need to attend to compositional data in the right manner, which subject is considered further on in this chapter), and to scan the data for atypical observations, etc. In this latter respect, it can be useful to precede a canonical correlation analysis by scrutinizing the data-matrix, using Krzanowski's (1987a,b) methods for cross-validational principal components to ferret out atypical and influential observations. The analysis can then be repeated with such observations removed from the data-matrix and the results compared with the original analysis.

The vector correlation coefficient

People often ask if there exists a way of expressing the correlation between sets of variables by means of a single number, given the interpretational problems attaching to canonical correlation coefficients. Escoufier (1973) proposed a generalization of Pearson's "coefficient of determination", the square of the correlation coefficient, which he called the coefficient of *vector correlation*, and which he denoted RV.

It is simply defined as

$$RV = \frac{Trace(\mathbf{S}_{12}\mathbf{S}_{21})}{\sqrt{Trace(\mathbf{S}_{11}^2)Trace(\mathbf{S}_{22}^2)}} \qquad (6:1)$$

This statistic is output by the program provided. As one might expect, Escoufier's coefficient ranges between 0 and 1.

The program provided here is called *cancorr.exe*. The input details for using it now follow:

Instructions for using the program **cancorr**

Line 1: Title of the job

Line 2: 1 number of variables in larger set
2 number of variables in smaller set
3 size of the sample
4 1 for a set of principal component analyses
0 is default
5 1 if logarithms are to be taken
0 is default
6 1 for multiple correlations, regression coefficients, partial correlations
0 is default (only available if the canonical correlations are computed for the correlation matrix).
7 1 for input as a processed covariance or correlation matrix. Type 0 for input as a data matrix
8 1 for a listing of the data matrix, 0 for no listing
9 1 if the covariance matrix is to be analysed, 0 if the correlation matrix is to be used (data-matrix input).

Line 3 and following: the data.

Output details

The program computes standard univariate statistics, multiple correlations, multiple regressions (for the correlations option), and produces scores, placed in the file *ccrplot* for subsequent insertion into *Graph Server*. The main output presents the full set of canonical correlations together with the redundancy equivalents. Normally, the redundancy analysis will only be really useful in the case of the correlation option.

A simple example now follows. It consist of three measurements, length, height and breadth, on the carapace of an Eocene ostracod species of the genus *Echinocythereis* from Aragon, Spain and three chemical parts determined on the same samples, to wit, vanadium, bromium and chromium. The material derives from petroleum exploration by ELF (Pau), to which organization RAR expresses his thanks for access to this valuable material. The data file is called *canco2.dat*. Some of the output will now be commented upon. Note that the program also provides, if asked, a full set of partial and multiple correlations and multivariate regression coefficients. A set of principal component analyses of the correlation matrix and the submatrices can also be obtained by following the foregoing instructions. Look now at the results listed in **Box 19**.

Box 19: Example of canonical correlation analysis: the Spanish palaeoenvironmental data

Program: **cancorr**

Introduction

Data: **canco2.dat**

variables in left set = 3 (length, height and breadth of the ostracod carapace)
variables in right set = 3 (boron, chromium, vanadium)

(It is convenient to make the "left hand set" the one containing the greater number of variables: there is no special statistical significance involved.)

number of observations = 10

Variable	Mean	Standard deviation
1	0.264	0.1886
2	0.658	0.1821
3	0.075	0.0481
4	0.360	0.0789
5	0.424	0.0389
6	0.248	0.0385

Correlation matrix for all variables

```
1    1.0000
2   -0.9670    1.0000
3    0.2142   -0.0355    1.0000
4   -0.5147    0.3837    0.6058    1.0000
5    0.5546   -0.5490   -0.1008   -0.4270    1.0000
6   -0.0951    0.1308   -0.1798   -0.2047   -0.4830    1.0000
```

Correlations for the left-hand set

These values are for the shell-dimensions. There is a very high negative correlation between length and height of the carapace. The other correlations are very small.

```
1    1.0000
2   -0.9670    1.0000
3   -0.2142   -0.0355    1.0000
```

Correlations for right-hand set

```
1    1.0000
2   -0.4270    1.0000
3   -0.2047   -0.4830    1.0000
```

The correlations between the chemical factors are moderate to low.

Correlations between sets

```
1  −0.5147    0.5546   −0.0951
2   0.3837   −0.5490    0.1308
3   0.6058   −0.1008   −0.1798
```

This set of correlations is interesting in that it exposes several sources of likely interaction between shell-morphology and the chemical factors.

```
Latent Root   1   0.58702
Latent Root   2   0.21062
Latent Root   3   0.06503
```

Percentage for Latent Root 1 = 68.05
Percentage for Latent Root 2 = 24.41
Percentage for Latent Root 3 = 7.54

Latent vectors

	Number 1	Number 2	Number 3
1	0.8155	0.8750	0.4931
2	−0.3108	1.3233	0.4940
3	−0.1322	0.4419	1.2539

Squared canonical correlation 1 = 0.587
Canonical correlation 1 = 0.766

Coefficients for the right set

0.8155   −0.3108   −0.1322

Coefficients for the left set

5.2845   5.7917   2.0544

Scores for canonical correlation 1

The scores are saved in file *ccrplot* for plotting. The results for plotting are listed below:

Specimen	Left-hand scores	Right-hand scores
1	−1.8319	−1.7508
2	−0.8613	−0.6072
3	0.2173	−0.1462
4	1.0349	0.0814
5	1.2008	1.6321
6	−0.3354	0.7699
7	−0.0726	−0.6958
8	0.4115	−0.4240
9	−0.8884	−0.1250
10	1.1251	1.2660

Redundancy analysis for correlations between original variables and new canonical variables

Correlations between original left-hand set and the new variables

−0.7563   0.6085   0.7167

Correlations between original right-hand set and the new variables

0.9753   −0.5951   −0.1491

Note: It is often found that these two vectors give a more understandable reification of a canonical correlation analysis using the correlation matrix as starting point than the canonical variate vectors, from which they are computed in a manner reminiscent of the techniques of factor analysis. A more striking example occurs in Chapter 7.

Proportion of variance of left set
explained by canonical correlation 1 = 0.4853

Proportion of variance of left-hand set explained by
canonical correlation 1 of the right set = 0.2849

Proportion of variance of right set
explained by canonical correlation 1 = 0.4425

Proportion of variance of right-hand set explained by
canonical correlation 1 of the left set = 0.2598

Note: This lopsidedness in variance partitions is an outcome of the "asymmetry property" of redundancy analysis.

Vector of correlations between original left-set variables and canonical variates of right set variables

−0.5795   0.4662   0.5491

Vector of correlations between original variables and canonical variates of left-set variables

0.7472   −0.4559   −0.1142

The canonical correlation scores can provide useful visual information about the correlation structure linear relationships of a sample.

Discussion of the canonical correlation analysis

This simple little example presents several points of importance for understanding the way in which a canonical correlation analysis appears in practical work. The directly computed canonical vectors differ rather strongly from the corresponding redundancy vectors due no doubt to the fact that the data do not fit the multivariate normal distribution very well and the sample-size is small. Both do, however support the view that the first component of each vector is important in establishing the observed correlation. The example also brings out a sad fact of geostatistical life, to wit, that geological data usually deviate from theoretical nicety for which reason it can sometimes be very difficult to get much sense out of the canonical correlation analysis of a set of analyses. A further training set is, which illustrates the capabilities of the program. This is in the file *canco4.dat*.

The method of *redundancy analysis* is an attempt by Stewart and Love (1968) to clarify the interpretation of the results yielded by canonical correlation analysis. The redundancy is defined as the proportion of the variance extracted by the canonical factor R_{dx} multiplied by the proportion of shared variance between the factor and the corresponding canonical factor of the other set. It expresses the amount of overlap between the two sets that is contained in the first canonical relationship, and so on (Cooley and Lohnes, 1971, p. 170). The feeling that one is not doing something quite "kosher" with all the seemingly subjective juggling with matrix multiplications, has created a good deal of hesitancy on the part of many practitioners in adopting the method. Gleason (1976) has, however, shown that the procedure is statistically, and mathematically, correct. Providing the data are not too different from the multivariate normal condition, the results yielded by the unembellished canonical correlation computations should prove satisfactory for most geological purposes. There is a geological example in Reyment (1991, p. 68) in which the redundancy vectors are reified to a scientifically appealing result and an analysis of the occurrence of lead and zinc in an earlier article (Reyment, 1972). The method has not caught on very well and Jackson (1991), for example, noted it as well as its competitors, but did not exemplify the procedure.

LOG-CONTRAST CANONICAL CORRELATION ANALYSIS

Among the many exciting innovations made by Aitchison (1986), we find a version of compositional canonical correlation analysis. This adaptation of canonical correlation analysis for use with compositional data is a moderately uncomplicated substitution of the usual covariances for those of the centred log-ratio covariance matrix. Hence, the *log-contrast canonical correlation* between a *C*-part composition x_1 and a *D*-part composition x_2 is computed in the usual manner, using the appropriately partitioned log-ratio covariance matrix Σ. It is defined as the maximum correlation between two log-contrasts $a_1^T \log x_1$ and $a_2^T \log x_2$ (known as the *log-contrast canonical components* when standardized by the appropriate unit variance constraints). Thus, the method of determination is

$$(\Sigma_{21}\Sigma_{11}^{-1}\Sigma_{12} - \lambda^2 \Sigma_{22})b_2 = 0 \tag{6:2}$$

from which

$$b_1 = \lambda^{-1}\Sigma_{11}^{-1}\Sigma_{12}b_2. \tag{6:3}$$

For present purposes, we use the arguments developed by Aitchison (1986, pp. 190–191) to adopt also the symmetric version by which the centred log-ratio covariance matrix replaces the log-ratio covariance matrix of (6:2) and (6:3).

Instructions for using the program **ccrconst**

This program is a suitably modified version of *cancorr.exe* for computing compositional log-contrast canonical correlations by means of the centred log-ratio covariance matrix, respectively, the associated log-ratio correlation matrix. There will therefore be a full complement of parts in the analysis. Following Aitchison's (1986, pp. 203, 329–331) implied recommendation, we suggest the computations be made on the covariances although there are many cases in which the log-ratio correlation version can provide interesting information. Anderson (1984) gives a trim account of the connection between canonical correlations obtained by covariances in relation to those derived from the correlations.

Input details

Line 1: The dimensionality of the problem; hence, if there are *m1* parts in the first set and *m2* parts in the second set, the dimensionality supplied here is the sum of these.

Line 2: Title of the job

Line 3: 1 Number of parts in the "left" set
         2 Number of parts in the "right" set
         3 Total number of specimens
         4 Type 1 for log-ratio covariances, or 0 for log-ratio correlations
         5 Type 1, if a set of principal component analyses is desired, else put = 0
         6 Type 1, if the input data-matrix is to be listed, else type 0

Line 4: The data-matrix

The coordinates for plotting in *Graph Server* are stored in *concrplt*.

As an example of the output, we shall consider the brief analysis of data on the composition of bentonites in the file *bentccr.dat,* and displayed in **Box 20**.

Box 20: Chemical Composition of Bentonites

Program: **ccrconst**

Data: **bentccr.dat**

Total number of parts = 11

Our aim is to compute canonical correlations for centred log-ratio data with 9 left-side parts and 2 right-side parts forming the bentonite data, taken come from an article by Cadrin et al. (1996) on isotopic and chemical compositions of bentonite as palaeoenvironmental indicators of the Cretaceous Western Interior U.S.A.

Variables in left set = 9

These are the parts Si, Al, Fe, Mn, Mg, Ca, K, Na and H_2O.

Variables in right set = 2

These are two determinations on oxygen isotopes.

Number of observations = 14

Variable	Mean	Standard deviation
1	2.954	0.2191
2	1.907	0.2386
3	−0.749	0.7764
4	−6.172	2.0926
5	0.327	0.3021
6	−0.705	0.8185
7	−1.911	0.6720
8	−2.415	0.9116
9	1.096	0.2477
10	3.621	0.2027
11	2.045	0.3069

We note here that part 4 is out of phase with the other parts in that its standard deviation is much greater than any of the others. This suggests the possibility of an atypical observation in the data-set.

Wilk's test of canonical correlations

Lambda one = 0.0001

Chi-square = 73.62

Degrees of freedom = 18.

Lambda two = 0.200

Chi-square = 13.21

Degrees of freedom = 8.

The tests indicate one definitely significant canonical root and a possible second significant root.

Squared log-contrast canonical correlation 1 = 0.9995

Canonical Correlation 1 = 0.9997

The first canonical correlation is almost 1 and is, therefore, very high indeed (canonical correlations run from 0 to 1).

Coefficients for right set

0.5164   0.5164

Coefficients for left set

−0.1357   −0.5247   −1.3975   −3.7639   −0.5915   −1.4554
−1.1580   −1.6744   −0.4568

The scores are preserved in the file *concrplt* for subsequent plotting.

MULTIPLE REGRESSION

Multiple regression may be conveniently regarded as a special case of canonical correlation where one of the sets consists of a single variable (Draper and Smith, 1966). The program for computing canonical correlation produces a multiple regression analysis by default if one of the sets contains just one variable; however, granted that parts of the output may be difficult to interpret, we have provided a simple procedure for computing multiple regression. This program is called *multregr.exe*.

Instructions for using the program **multregr**

Input details

Line 1: Title of the job.

Line 2: 1 Number of variables
        2 Number of observations

Line 3: The data-matrix in free format; note, that the dependent variable must be placed last – i.e., the last column of the data matrix.

Output details

Means, standard deviations, covariances, correlations, multiple correlation coefficient, regression coefficients and the linear intercept of the regression. The output also includes a comparison between the values of the dependent variable

Multiple regression

and those computed by the regression and, hence, the residuals resulting from the multiple regression, that is, the difference between each observed value of the dependent variable and the estimated value of it, obtained from the multiple regression equation. This is useful information because it lets you know how well your regression fits the data and, moreover, points to individuals that deviate from the main body of the material. A simple example is now presented in **Box 21**. It concerns a data set of 38 specimens of the three mid-Cretaceous ammonite species *Neogastroplites americanus, N. cornutus* and *N. muelleri* from the Cenomanian of northwestern USA and five variables. The independent variables are four determinants of shape (obtained by Bookstein's (1991) method of relative warps) and the dependent variable is counts on rib frequencies.

Box 21: Multiple regression analysis of *Neogastroplites* spp. Three Cretaceous ammonite species from the Cenomanian of North America.

Number of variables = 5; observations = 38

Correlation matrix

	1	2	3	4	5
1	1.0000	−0.1979	−0.1055	0.0310	0.5253
2	−0.1979	1.0000	0.2736	−0.0379	−0.1981
3	−0.1055	0.2736	1.0000	0.3135	−0.1518
4	0.0310	−0.0379	0.3135	1.0000	−0.0037
5	0.5253	−0.1981	−0.1518	−0.0037	1.0000

The first $m - 1$ variables are the predictor (independent) variables.

The value of multiple R square is = 0.2907, the square root of which yields the multiple correlation coefficient, which is 0.5392.

$F_{4,33}$ for ANOVA on the multiple *R*-value is 3.3818, which is significant on the 5% level.

The beta weights (coefficients) for the multiple regression equation are:

1	27.9615
2	−2.9039
3	−2.3470
4	0.0595

The associated intercept constant is 21.947.

Observed dependent fitted response residuals variable (the lateral ribs)

```
21.000     19.898     1.102
. . . . . . . . . . . . . . . . . . . . . . . . . . . . . . .
33.000     27.144     5.856
21.000     19.898     1.102
```

There are two very poor fits. The analyst would be advised to check the associated specimens for clerking errors and, or, deformation of the specimens. In a full analysis, one would not stop here, the next step being to perform graphical studies. Various plots of residuals can help model improvement. Useful guides for doing this are Atkinson (1985) and Cook & Weisberg (1982).

As an exercise, see what you can make of the data-set *multregr.dat*. This consists of seven distance-measures determined on the carapaces of 70 individuals of the ostracod species *Veenia rotunda* (Cretaceous, Morocco). A subset of these data appeared in **Box 18**. The characters are length of carapace, maximum height over the anterodorsal angle, distance from the eye tubercle to the posteroventral corner, maximum height over the posterodorsal angle, distances of the adductor tubercle from the anterior and ventral margins and the length of the posterior margin of the carapace.

SUMMARY

1. Canonical correlation analysis.
2. Redundancy analytical model.
3. Escoufier's vector correlation coefficient.
4. Log-contrast canonical correlation analysis.
5. Multiple regression.
6. Programs and associated training sets
   cancorr   canco2.dat
   ccrconst   bentccr.dat
   multregr   multregr.dat

Chapter 7

Some problems in petrology and geochemistry

INTRODUCTION

It should by now be apparent to you that multivariate data abound in everyday geology, particularly geochemistry, petrology and palaeontology. In fact, if a statistical survey were to be carried out it would probably be found that most multivariate situations in geology (excluding palaeontology) arise in the study of the compositions of rocks of various kinds. It is, however, here that most errors in statistical applications occur, as has been well exposed by Aitchison (1986) in his fundamental monograph.

THE SAUDI-ARABIAN RIFT VALLEY VOLCANICS

The examples presented in this chapter are rather typical of their genre. We shall begin with a study of volcanic rocks. Demange et al. (1983) studied a recent N–S trending volcanic chain, the Jabal al Abyad, located along a fractural axis, classified by those authors as a rift-valley, the rocks of which have evolved from mildly alkaline basalts to phonolites under certain structural conditions, and to comendites under others. The analyses of the chemical data were made using simple graphs of ratios, the statistical soundness of which type of procedure is not unchallengeable. Chemical and mineralogical variations were explained as manifestations of fractional crystallization. There is an unexpected complication to understanding the importance of a major part of the publication in that there is no explanation of the symbols used in the graphs (the Fig. 4 of the authors lacks the key to which that figure and several others refer).

The data listed in the file *jabal.dta* consist of determinations made on 11 samples of the major elements, expressed as oxides, SiO_2, TiO_2, Al_2O_3, Fe_2O_3, FeO, MnO, MgO, CaO, Na_2O, K_2O, P_2O_5 and an unspecified residue. The trace elements assessed were V, Cr, Co, Ni, Cu, Zn, Li, Rb, Sr, and Ba. There are 11 major elements and 10 trace elements in the data-matrix. It is quite obvious that the first set of determinations consists of parts which sum to a constant. What is the situation for the trace elements? Although there does not seem to be an obvious constraint involved, there is, however, indeed. These data are expressed as parts per million and are no more or no less than the "parts" we met in the first chapter.

Although Demange et al. (1983) did not compute any statistics for their data, other than a very few simple univariate ones, they did perform some graphical operations having a statistical import. We need only to consider the use of ratios. It is rather common praxis in petrology, it seems, to attempt to arrive at meaningful ordinations of data by plotting ratios that are interpreted as possessing special diagnostic significance. One such graph used in the study examined here is the ratio Ca/Sr plotted against Sr. Another is the graph of the ratio K/Ba plotted against Ba. The statistical, and logical, objection that comes to the fore here is that the same component enters into both axes of the graph; strontium is being compared with itself and barium likewise with itself. This is a questionable procedure at best and was decried by Karl Pearson more than a hundred years ago. It is also difficult to desist from concluding that some aspects of petrological classification bear the stamp of arbitrariness and the names applied to rock compositions depend on time, place, person and chance (see also Sørensen, 1974).

Suggested analytical approach

The principal value of the data-set lies with the possibility it offers of finding natural groupings in the samples, and hence, the rock-types subjected to chemical analysis. The first problem to be overcome is that there are more variables than samples, namely, 21 chemical parts and only 11 samples. This indicates that a Q-mode strategy is going to be required, using a rather special arrangement of the data with the variables taking the role usually played by the specimens. It is, as it were, as though we had inverted the space occupied by the data. A good method with which to begin is that of *principal coordinates*. We shall apply the method to the raw data-matrix and then to its log-ratio equivalent as a start.

The second area of interest centres about the relationship between major elements and trace elements. *Canonical correlation analysis* supplies a possible means of assessing this but, owing to the paucity of observations available, it will be necessary to cull the number of variables so as not to invalidate the covariance matrix. In reducing the dimensionality of the problem, we have relied on the elements betokened as being diagnostic in the article by Demange and his coworkers.

Principal coordinate analyses

The salient results obtained by applying the program *pcoord.exe* to various versions of the data-matrix for the samples of igneous rocks from the Saudi rift-valley (Demange et al., 1983) are displayed in **Boxes 22** and **23**. It is a good example of the power of Q-mode ordination for resolving a difficult problem. The first box contains an abridged version of the results obtained for the computations applied to the raw data-matrix. The second of these boxes presents a summary of the same computations implemented for the log-ratio data-matrix.

Box 22: Principal coordinate analysis of the raw data-matrix for the Saudi rift-valley data

(a) The data-matrix in terms of the 11 rock-samples
    Data in file: *jabal11.dat*

Data from Demange et al. (1983) for the oxides SiO_2, TiO_2, Al_2O_3, Fe_2O_3, FeO, MnO, MgO, CaO, Na_2O, K_2O, P_2O_5

The minimum spanning tree

Connected to	by	the distance
hawaiite_1	basalt	1.5929
hawaiite_2	hawaiite_1	2.5447
mugearite	hawaiite_2	1.3812
benmoreite	mugearite	3.6931
quartz trachyte	benmoreite	2.3830
trachyphonolite	quartz trachyte	1.2421
alkaline phonolite	trachyphonolite	1.3468
phonolite	alkaline phonolite	1.3999
comend. trachyte	alkaline phonolite	1.5294
comendite	comend. trachyte	0.7943

Length of tree = 17.91.

The minimum spanning tree is a useful, almost essential, tool for making an ordination comprehensible. Points that may seem to be close neighbours in the bivariate scatter plot may well be really quite distant from each other, particularly if, say, the third latent root is almost as large as the second latent root (and hence represents important spatial information not accessible in a bivariate plot).

Latent roots (largest three) of transformed association matrix

1.9010
0.5767
0.3747

Specimen	Coordinates		
1	0.6113	−0.3959	0.0591
2	0.5505	−0.2772	0.0122
3	0.5192	0.2356	−0.2369
4	0.4226	0.4131	−0.1033
5	0.0172	0.1759	0.5023
6	−0.2689	0.1725	−0.0353
7	−0.3612	−0.0180	0.0114
8	−0.4007	0.0037	−0.0745
9	−0.2354	0.0189	0.1158
10	−0.4269	−0.1392	−0.1239
11	−0.4276	−0.1894	−0.1269

These points are saved in the file *prcrd* for plotting by means of *Graph Server*.

Values of residuals

Roots exceeding	Percentage residual
2	35.42
3	25.68

Discussion of results in **Box 22**

The minimum spanning tree, superimposed on the plot of the first two principal coordinates, gives a clear indication of progression in chemical properties as a function of time from basaltic rocks to phonolitic in reasonable accordance with the deductions of Demange et al. (1983). There is a high degree of integration in the data which explains the quite low residuals obtained. This condition is a favourable one in a Q-mode analysis, since it preserves distances between objects. Its statistical effects are further manifested in the concluding section of this chapter.

We now pass to the principal coordinate analysis of the log-ratio data-matrix of oxides, summarized in **Box 23**.

Box 23: Principal coordinate analysis of the constrained data-matrix for the Saudi rift-valley

The Saudi rock sequence in log-ratio principal coordinate analysis using program *pcrdcons*.

Data in file: *jabal111.dat*

Variables = 11 Individuals = 11

	Minimum spanning tree	
Connected to	by	distance
hawaiite_1	basalt	1.1414
hawaiite_2	hawaiite_1	1.8417
mugearite	hawaiite_2	0.6344
benmoreite	mugearite	2.9021
quartzose trachyte	benmoreite	2.3247
trachyphonolite	quartzose trachyte	1.0122
comenditic trachyte	trachyphonolite	1.8021
phonolite	comenditic trachyte	1.5801
hyperalk. phonolite	comenditic trachyte	1.6877
comendite	hyperalk. phonolite	2.1694

Length of tree = 17.096

Latent roots (largest three) of transformed association matrix

1.782
0.476
0.311

Specimen	Coordinates		
basalt	−0.5181	0.1877	−0.2161
hawaiite_1	−0.5179	0.1227	−0.2188
hawaiite_2	−0.4636	0.1008	0.2734
mugearite	−0.4010	0.0202	0.3130
benmoreite	−0.1500	−0.3496	−0.1836
trachyphonolite	0.1651	−0.2301	0.0287
phonolite	0.3568	−0.1278	0.0537
hyperalk. phonolite	0.4656	0.2005	0.0203
quartzose trachyte	0.1096	−0.2549	−0.0164
comenditic trachyte	0.4092	−0.0135	0.0190
comendite	0.5441	0.3438	−0.0733

These points are saved in the file *prcrdplt* for plotting.

Values of residuals

Roots exceeding	Percentage residual
2	35.71
3	26.85

These residuals are relatively small and it may therefore be concluded that the principal coordinate solution fits the data well.

Discussion of the constrained principal coordinate analysis

Although the general form of the results seems to be much the same as was obtained with the raw data-matrix, and the relatively small residuals are closely comparable, there are, nonetheless, some significant deviations that are worth noting. If you examine the plot of the points in the plane of the first two coordinates you will observe that the first two thirds of the path traced out by the minimum spanning tree (constructed from the log-ratio data) is the same up to the location of the sample of trachyphonolite. Thereafter, the ordering of the samples differs, to end up, however, in both cases with comendite. The ordination for the log-contrast principal coordinates, with the minimum spanning tree superimposed, is shown in Fig. 32.

Whichever of the two results makes the best scientific sense requires expertise in petrological interpretation. The salient feature of the present analysis is that the statistically correct model leads to a somewhat different outcome from that yielded by the unsophisticated one. However, in both cases, the high correlations occurring between variables guarantee that the distances between objects are well preserved in the plane of the first two principal coordinates.

Discussion of the Saudi Rift data

The foregoing set of analyses indicates that the main conclusions arrived at by Demange et al. (1983) are supported by the multivariate analysis, albeit with certain minor reservations. Even a cursory perusal of the input material discloses that there are very pronounced trends in the observations and it is therefore not surprising that such obvious numerical structure shows up in almost any kind of quantitative appraisal. At the statistical level, the results achieved with the aid of the appropriate model in terms of compositional parts do not greatly differ from the inappropriate model for the Q-mode analysis. The treatment of the data summarized in the present chapter can be expanded so as to include a correspondence analysis and analyses of subsets (i.e. appropriately constructed subcompositions) of the data.

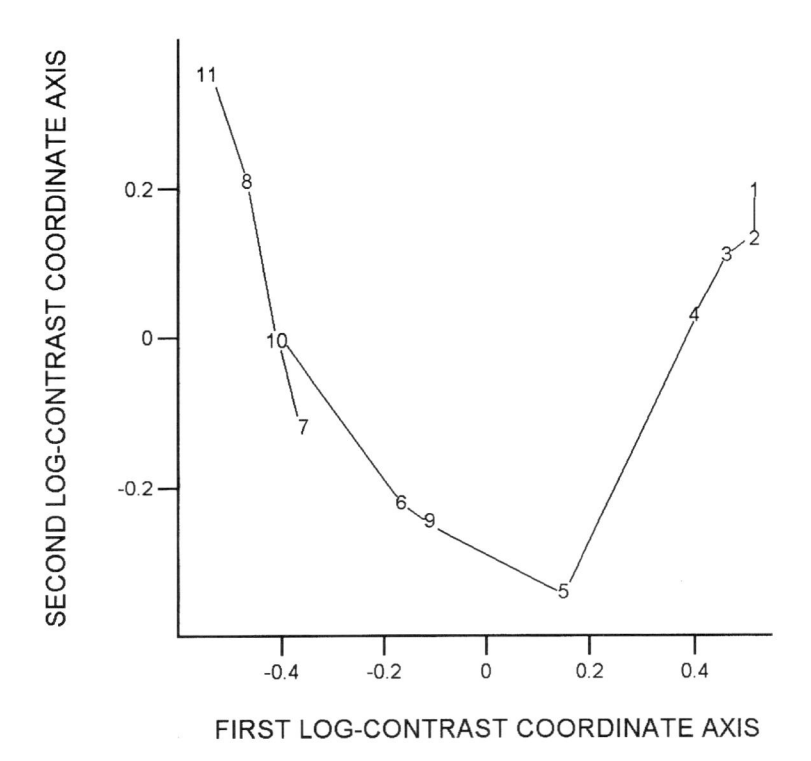

Fig. 32. Ordination of alkaline rock categories by log-contrast principal coordinate analysis with the minimum spanning tree superimposed on the plot. KEY: 1 Basalt; 2 Hawaiite-1; 3 Hawaiite-2; 4 Mugearite; 5 Benmoreite; 6 Trachyphonolite; 7 Phonolite; 8 Hyperalkaline Phonolite; 9 Quartzose Trachyte; 10 Comenditic Trachyte; 11 Comendite.

A PALAEO-OCEANOGRAPHICAL EXAMPLE

Introduction

Cadrin et al. (1996) studied the isotopic and chemical compositions of bentonites as possible palaeo-environmental indicators of the Cretaceous Western Interior Seaway of Northern America. The bentonites were collected from the Cenomanian-Turonian of the Greenhorn cycle and the Campanian of the Claggett sequence. A principal objective of their investigation was to attempt to find out why oxygen and hydrogen isotopic compositions of the volcanic ash do not always agree with those inferred for the seas themselves, as assessed from well preserved fossils. The deviations from theory were attributed to alteration of the bentonites after they had been deposited.

The Cretaceous Period was a time of great episodic epicontinental flooding, the most important of which took place in the Cenomanian to Turonian. This is one of the mondially registered tectono-eustatic, or Suessian type, transgressions. The next in time, though generally less extensive, was that of the Campanian-Maastrichtian. The issue taken up by Cadrin et al. (1996) is one of practical significance in palaeogeography in that chemical constituents of sediments can be expected to provide evidence of past oceanographical conditions.

The Data

In the following, the chemical analyses listed in Tables 1 and 2 of Cadrin et al. (1996) are analysed by two multivariate methods. These are log-contrast principal component analyses of the oxides of major elements, using the centred log-ratio covariance matrix and constrained canonical correlation analysis of the oxides with the two isotopes, using the log-ratio covariance matrix. The major element oxides are SiO_2, Al_2O_3, Fe_2O_3, MnO, MgO, CaO, K_2O, and H_2O, all expressed as weight-percent oxide and hence compositional. The two isotopes are δD in ppm and δO in ppm. The published analytical data does not sum exactly to 100 and it must be assumed that some components were not included in the printed table.

There are four samples of the Cenomanian marker bed, known as the "X-bentonite", from the Belle Fourche Member of the Big River Formation. This marker can be traced over two thirds of the Western Interior Basin. Notwithstanding the synchrony of this ash-fall, the occurrences display considerable variation in mineralogy, a condition that is ascribed by many workers to post-depositional processes. Samples 5 and 6 are from the Cenomanian of Manitoba. There are four Turonian samples, two each from Duck Mountain, respectively, Riding Mountain, Manitoba. Two Turonian samples come from Montana, from localities at Billings, respectively, Great Falls. The data also contain two samples of Campanian age, both from Riding Mountain, Manitoba.

The Problem

It is generally surmised that the northern part of the Western Interior Basin contained cool waters that moved southwards to meet warmer waters from the south at the instant of union of the two arms in Late Cenomanian time. It is reasonable to expect that there would have been variations in water chemistry. There is some uncertainty about how long it would take for a homogeneous chemical system to develop or, even, whether such a system could be expected to appear over the time during which the epicontinental sea was at a maximum. If, then, the appropriate ordinations of sampling sites indicate the likelihood of randomness in the distributions of points obtained by the usual procedures, it may be inferred that homogeneous conditions were quick to become established. The opposite result would speak for continued heterogeneity in marine conditions. Cadrin et al. (1996)

made no statistical analyses of any of their data; however, a bivariate plot of the oxygen and hydrogen isotopes was taken to indicate a difference to exist between the Cenomanian and Turonian material.

Log-contrast principal component analysis

The results of the log-contrast principal component analysis are shown in **Box 24**.

Box 24: Principal component analysis of eight major oxides; the North American Cretaceous bentonite data.

Although we have said this before, we repeat it here again: you cannot set great store by correlations in simplex space and there is probably little to gain by a 'one-to-one' comparison of simplex and raw correlation coefficients. Aitchison (1997) makes this point quite clear.

Latent roots

3.9190 1.5653 1.2006 0.8188 0.3490 0.1302 0.0173

Percentages of the total variance (trace)

Latent Root 1 = 48.98
Latent Root 2 = 19.57
Latent Root 3 = 15.00
Latent Root 4 = 10.23
Latent Root 5 = 4.36
Latent Root 6 = 1.63
Latent Root 7 = 0.22

Latent vectors

	1	2	3	4	5	6	7
1	0.4715	0.1628	0.1826	0.1610	0.1205	−0.2895	−0.7559
2	0.4551	0.1738	0.1789	0.1373	−0.3779	−0.4962	0.5508
3	−0.1720	0.5699	−0.3891	0.4752	0.1275	0.1533	0.0769
4	0.3956	−0.2749	0.2536	0.1557	0.6730	0.2946	0.3156
5	0.1772	−0.4975	−0.5976	−0.2598	−0.0851	−0.1484	−0.0520
6	−0.2705	−0.4103	0.4394	0.4316	−0.4168	0.1836	−0.1178
7	−0.3264	0.2444	0.4070	−0.5763	0.1926	−0.1969	0.0307
8	0.4162	0.2591	0.0547	−0.3450	−0.3939	0.6819	−0.0444

We venture here, as a didactic exercise only, a reification of the latent vectors. The important contributors to the first latent vector are Si, Al, Mg and water. The second latent vector represents significant responses from Fe, Ca, and K. The third vector contains significant loadings for Fe, Ca, K and Na. The "invariant latent vector", the seventh, represents covariation in Si and Al (termed "invariant" because it corresponds to the smallest latent root and, hence, that linear combination of the parts that expresses the least variability in the data-set). The complete output for this problem is obtained from analysing the data in the file **bentcon.dat**. Note that part 4 gives very high variances and covariances owing to the fact that we were obliged to use the 'add-a-minute value' to several of the observational vectors because of the virtual redundancy of that part. You may perhaps wish to compare the values for the constrained and "usual" latent vectors. The points for plotting are saved in the file *aitchplt*.

Log-contrast canonical correlation analysis

Aitchison (1986) proposed an adaption of classical canonical correlation analysis for use with compositional data; this topic was reviewed in Chapter 6. The canonical correlation between a *c*-part composition and a *d*-part composition is computed in the usual manner from the appropriately partitioned log-ratio covariance matrix. In Chapter 6, the centred log-ratio covariance matrix was used, which is Aitchison's preference. By way of contrast, we here invoke the log-ratio covariances. This was done by combining the program for computing standard canonical correlations **cancorr.exe** with the appropriate log-ratio data-matrix for the major oxides.

As pointed out above, the appropriate model for constrained data in a canonical correlation analysis is that obtained by the log-ratio variances and covariances. As defined on p. 219, the analysis is therefore in terms of log-contrasts. SiO_2 was selected as the common divisor, leaving thus seven components for the major oxides. The two right-hand variables are the content of the isotopes of hydrogen and oxygen.

Our aim is to ascertain to what degree the oxides are correlated with the isotopes. The data are in the file **bentccr.dat.**

The log-ratio canonical correlations for Cadrin's bentonites

Number of observations = 14

A palaeo-oceanographical example

Variable	Mean	Standard deviation
1	−1.048	0.0787
2	−3.703	0.7995
3	−2.627	0.2021
4	−3.659	0.8118
5	−4.865	0.7438
6	−5.369	0.9175
7	−1.859	0.1488
8	1.081	0.0941
9	0.226	0.0381

The oxides are entries 1–7 and the isotopes, entries 8–9.

Array for covariances of the left-set (the oxides)

```
1    0.0062
2    0.0069    0.6392
3   −0.0069   −0.0859    0.0408
4    0.0069   −0.1459    0.0525    0.6590
5    0.0192    0.0460    0.0382   −0.0170   0.5532
6    0.0137    0.1912   −0.0341   −0.2255   0.2878   0.8418
7    0.0056    0.0222   −0.0077    0.0301   0.0125   0.0675   0.0222
```

Array for covariances of the right-set (the isotopes)
```
1    0.0089
2    0.0018    0.0015
```

Array for the covariances between sets
```
1   −0.0036   −0.0016
2   −0.0124   −0.0159
3    0.0032    0.0016
4   −0.0115    0.0061
5   −0.0054   −0.0182
6   −0.0515   −0.0246
7   −0.0110   −0.0025
```

Weierstrass Diagonalization

Latent Root 1 = 0.9326

Latent Root 2 = 0.6797

Percentage for latent root 1 = 57.84

Percentage for latent root 2 = 42.16

Wilk's lambda-test of canonical roots

Lambda one = 0.0216

Chi-square = 34.518

Degrees of freedom = 14. The significance level is 23.40 for 14 degrees of freedom and the canonical correlation is statistically secure.

Lambda two = 0.3203

Chi-square = 10.247. For 6 degrees of freedom, this value is not statistically significant.

Canonical Correlation 1 = 0.9657

Hotelling canonical coefficients for right set (the isotopes)

4.73   18.27

Hotelling canonical coefficients for left set (the oxides)

−8.44   −0.59   −2.61   0.08   0.22   −0.66   −1.01

These values indicate that the between-sets correlation is largely due to the oxygen isotope with the log-ratio of the aluminium oxide and water.

Scores for canonical correlation 1

Specimen	Left-hand scores	Right-hand scores
1	19.802	6.873
2	21.068	8.360
3	21.209	8.407
4	21.505	9.232
5	21.424	9.224
6	22.503	9.804
7	22.520	9.826
8	22.837	9.899
9	23.088	10.772
10	23.703	10.615
11	22.371	9.490

12	22.070	9.494
13	21.384	8.745
14	21.403	8.767

Stewart-Love redundancy analysis

Correlations between original LEFT-set variables and the new left-side canonical variates

−0.0482   **−0.3608**   0.0455   0.0582   **−0.3706**   **−0.7177**   −0.1021

Correlations between original RIGHT-set variables and the new right-side canonical variates

0.0753   0.0352

The redundancy analysis places more emphasis on the log-ratios of the oxides of Fe, Ca and K.
  Correlations between original variables of one set with the canonical variates of the other set.
  Correlations between original LEFT-set variables and canonical variates of RIGHT set

−0.0466   **−0.3484**   0.0440   0.0562   **−0.3578**   **−0.6931**   −0.0986

Correlations between original RIGHT-set variables and canonical variates of LEFT-set

0.0728   0.0340

The indication of these relationships is that the two isotopes are about equally as important in the canonical correlation with mainly the log-ratios of Al, Ca and K as important contributing major oxides. It is important to remember that these relationships are not necessarily the most important part of the correlation. They merely signify that there is a statistically significant combination of the above form in the data.
  The figure yielded by the plot of the canonical correlation scores for the first correlation produces a vague subdivision into Cenomanian and Turonian fields. The Campanian values are here, as elsewhere, anomalously situated.

Conclusions: The results of the two multivariate analyses under the log-contrasted model suggest that there is a slight indication of time-controlled discrimination in the distributions of sample values for the Cenomanian and Turonian, but this indication is by no means unequivocal in that there is overlap between the two. The Campanian samples fail to distinguish themselves as being compositionally different. The data-set of just 14 samples is hardly sufficient to allow of any definite conclusions, but the multivariate results obtained cannot be claimed to yield strong support for the time-oriented shift in properties suggested by Cadrin et al. (1996).

SUMMARY

1   The Saudi-Arabian rift-valley.
     (a) Data-files: jabal.dta, jabal11.dat, jabal111.dat.
     (b) Programs employed: pcoord, pcrdcons, cancorr, logcov.
     (c) Minimum spanning tree exemplified.
     (d) Methods: principal coordinates, canonical correlations.
2   A palaeo-oceanographical study.
     (a) Data-files: bentcon.dat, bentccr.dat.
     (b) Programs employed: pcaconst, logcov, cancorr.
     (c) Methods: log-contrast principal components and log-contrast canonical correlations.

Chapter 8

Miscellaneous Examples

In this chapter we present a collection of miscellaneous examples with the intention of covering briefly a variety of the kinds of problem that can arise in geological work. The methods and results are no more than sketched out, our wish being that you try you hand at performing the full analyses, using the concepts already imparted to you. The first of these case histories concerns the analysis of the time-ordered ranking of species of Eocene ostracods (cf. Reyment, 1985).

EVOLUTION IN ECHINOCYTHEREIS (OSTRACODA, CRUSTACEA)

There are 102 sampling levels and the data consist of quinquevariate means on the classical morphological carapace traits of length and height of carapace, diagonal distance from the eye-tubercle to the posteroventral angle, posterior height, and the length of the posterior margin. The aim of the analysis is to see whether the time-ordered sequence is correlated with location in time on the basis of evolution in the dimensions of the carapace. The dependent variable is just the ordination according to stratigraphical position. The results are displayed in **Box 25**.

Box 25: Multiple regression study of the *Echinocythereis* data. An evolutionary sequence of ostracods from the Eocene of Aragon, Spain.

Program: **multregr**
Data: **echinreg.dat**

Number of variables = 6; observations = 102

Correlation matrix

	1	2	3	4	5	6
1	1.0000	0.9686	0.9863	0.9714	0.6991	0.8831
2	0.9686	1.0000	0.9629	0.9447	0.7326	−0.8695
3	0.9863	0.9629	1.0000	0.9666	0.7195	−0.8662
4	0.9714	0.9447	0.9666	1.0000	0.6877	−0.8401
5	0.6991	0.7326	0.7195	0.6877	1.0000	−0.6314
6	−0.8831	−0.8695	−0.8662	−0.8401	−0.6314	1.0000

The multiple $R^2 = 0.790$, the square root of which yields the value of multiple correlation coefficient, which is $= 0.889$. $F_{5,96}$ for ANOVA on the multiple correlation coefficient is 72.23, which is highly significant.

Regression equation – predictors

Beta weights (the coefficients) for the multiple regression equation

1	−311.01
2	−142.00
3	66.01
4	143.74
5	28.55

Intercept constant 280.54.

Discussion of Box 25

Firstly, we note that the time-constituent is highly negatively correlated with the carapace dimensions. Inasmuch as these dimensions are largely a measure of size, there is here an indication that the evolutionary sequence is characterized by a decrease in size of the carapace. The multiple correlation coefficient is highly significant, which result provides support for the conclusion of size-reduction over time. The coefficients of the multiple regression equation are interesting in that they afford a glimpse into the morphological development accompanying the evolutionary process. This equation shows that most of the change in the carapace through time lies with the length and anterior and posterior maximum heights of the carapace. The shells seem to have become more "squarish" with time.

The results of the multiple regression provide a preliminary insight into a complex multidimensional relationship. In order to probe this relationship further, a principal component analysis may be performed on the sequence.

Principal component analysis of the ostracod data

The results of the principal component analysis of the 102 sequentially ordered levels of the species of *Echinocythereis* are displayed in **Box 26**. The same five variables as before, measures on traits of the carapace, were subjected to analysis by means of the program *pcomp2.exe*. The analysis was made on the covariance matrix and both simple principal components and principal component factor analysis were applied to the data in file *echinpca.dat*.

Box 26: Principal component analysis of the *Echinocythereis* data.

Program: **pcomp2**

Data: **echinpca.dat**

Covariance matrix

	1	2	3	4	5
1	0.0110	0.0054	0.0080	0.0065	0.0012
2	0.0054	0.0029	0.0040	0.0032	0.0006
3	0.0080	0.0040	0.0060	0.0048	0.0009
4	0.0065	0.0032	0.0048	0.0041	0.0007
5	0.0012	0.0006	0.0009	0.0007	0.0002

Correlation matrix

	1	2	3	4	5
1	1.0000	0.9686	0.9863	0.9714	0.6991
2	0.9686	1.0000	0.9629	0.9447	0.7326
3	0.9863	0.9629	1.0000	0.9666	0.7195
4	0.9714	0.9447	0.9666	1.0000	0.6877
5	0.6991	0.7326	0.7195	0.6877	1.0000

There are several very high correlations, and all are highly significantly different from nought. This is not an unusual situation for crustaceans in which the various dimensions of the carapace are often very highly integrated.

Latent vectors

0.02358   0.00020   0.00015   0.00012   0.00008

Corresponding percentages of the trace

97.681   0.837   0.634   0.515   0.333

The high correlations are accompanied by the natural outcome that almost all of the variation in the material is concentrated to the first latent root.

Latent vectors by columns

	1	2	3	4	5
1	0.6802	0.0442	−0.4781	0.2903	−0.4718
2	0.3402	0.5598	0.4775	0.4704	0.3485
3	0.5002	0.1197	−0.1712	−0.6880	0.4825
4	0.4074	−0.7482	0.5206	0.0515	0.0214
5	0.0735	0.3323	0.4930	−0.4674	−0.6501

The overwhelmingly dominant variational criterion is that of size, as is shown by the first principal component vector. The vector connected to the smallest latent root is also interesting in that it represents an almost invariant linear combination in the material.

Factor loadings matrix

	1	2	3	4	5
1	0.1045	0.0006	−0.0059	0.0032	−0.0042
2	0.0522	0.0080	0.0059	0.0052	0.0031
3	0.0768	0.0017	−0.0021	−0.0077	0.0043
4	0.0626	−0.0106	0.0064	0.0006	0.0002
5	0.0113	0.0047	0.0061	−0.0052	−0.0058

Confidence intervals for latent roots of covariance matrix

Bounds of confidence intervals for the latent roots

Lower bound for root 1 = 0.0185

Upper bound for root 1 = 0.0325

Lower bound for root 2 = 0.0001

Upper bound for root 2 = 0.0002

Lower bound for root 3 = 0.0001

Upper bound for root 3 = 0.0002

Lower bound for root 4 = 0.0001

Upper bound for root 4 = 0.0002

Lower bound for root 5 = 0.0001

Upper bound for root 5 = 0.0001

Confidence intervals can be computed in the case of the covariance matrix. There is, as yet, no corresponding theory for finding confidence intervals for the principal components obtained from the correlation matrix. An approximate solution can, however, be achieved by means of a jackknifing technique, such as we have exemplified for cross-validation principal component analysis (p. 123).

Principal component factor analysis with varimax rotation

A principal component factor analysis is, as the name implies, a principal component analysis embellished with certain factor-analytical appurtenances. It is not to be confused with "true" factor analysis of modern psychometrics (cf. Reyment and Jöreskog, 1993).

Varimax factor matrix

Var	1	2
1	0.0822	0.0645
2	0.0364	0.0383
3	0.0597	0.0484
4	0.0560	0.0299
5	0.0060	0.0106
Variance	0.30	0.18

Discussion of the principal component analysis

The principal component analysis yields further evidence for differentiation in size as a function of time of the lineage. The plot of the scores for the first two principal components shows there to be two groupings of the points. The plot of the varimax-rotated scores causes the groupings to appear in sharper relief. The logical next step in the analysis is to contrast these two groups by some suitable method, such as the discriminant function. The sequence under consideration is believed to have encompassed three species with *Echinocythereis isabenana* (Oertli) at the base, then *E. aragonensis* (Oertli) and, finally, *E. posterior* at the summit. For the purposes of the analysis displayed in **Box 28**, there are 46 observational vectors in the older sample and 56 in the younger one.

CARBONATES IN CI CHONDRITES: CLUES TO PARENT BODY EVOLUTION

Introduction

Endress and Bischoff (1996) were concerned with interpreting the history of formation and evolution of the parent bodies of a type of chondrites known as *CI chondrites* – other categories are CM, CO, CV, and CR chondrites. This branch of research on *meteorites* encompasses a very restricted body of data, notably six falls in the gm to kg range, of which only two exceed 1 kg in weight. These two, the falls at Alais and Orgueil, have been given most attention by various specialists and there has been a tendency to extrapolate results obtained for them to the other falls. The work of Endress and Bischoff was directed towards assessing the generality of this assumption. They used material from four of the meteorites, to wit, Orgueil, Ivuna, Alais and Tonk. The following case-history makes use of the published data in the article by Endress and Bischoff; however, examination of the scatter diagrams indicates that the sample sizes used for making these were greater than indicated by the tables of determinations – for example, the figures for Orgueil, which has meant that the treatment of the data could not be as complete as it undoubtedly merits.

CI chondrites are regolith breccias, consisting of various types of chemically and mineralogically distinct mineral and lithic fragments. They differ greatly with respect to degree of brecciation and intensity of aqueous alteration, which conditions appear to be linked to each other. The Ivuna material was inferred to incorporate four lithologies, two of which were identified in the Orgueil fall, although in different proportions. These lithologies are characterized by their mineralogies and by compositional differences in included dolomites (Endress and Bischoff stated that each lithology can be distinguished by the Fe and Mn contents of dolomites). Other characteristics were studied as well, such as the different origins of carbonate grains and fragments and the presence of the carbonates breunnerite, siderite and calcite. It was concluded that there were several episodes of alteration on the CI parent body, physicochemical conditions during carbonate formation were different among CI chondrites, and CI carbonates were formed at low temperatures in equilibrium with surrounding fluids. Lithology I is distinguished by a relatively high proportion of carbonates – up to 10%. Lithology II is typified by sulphates (pyrrhotites), lithology III contains phyllosilicate aggregates, and lithology IV is relatively rich in olivines.

Endress and Bischoff relied solely on plots of the data obtained from 18 thin-sections of the chondrites for assessing the eventual statistical properties of their results. Our present task is to see what can be extracted from the tabulated analyses by means of appropriate multivariate procedures.

The data

The components determined by Endress and Bischoff (1996) are CaO, MgO, MnO, and FeO. The tables also contain CO_2, which was obtained by subtraction from 100. There is, therefore, no doubt that we are dealing with compositional data. Moreover, the scatter diagrams in Fig. 10 of Endress' and Bischoff's article disclose the presence of atypical observations in some of the data. For the most part, however, the points are relatively tightly grouped. The chemical compositions of the carbonate minerals involved have a bearing on the interpretations of the results. Thus, dolomite is a calcium magnesium carbonate, breunnerite is a magnesium ferromanganese carbonate, and calcite is just calcium carbonate. The vast majority of the carbonate grains consist of dolomite, with breunnerite in second place. Calcite is reported to be rare.

We can expect spurious correlations to occur in the data, of the same kind as occur, for example, in tables of the A B O alleles of serology. The graphs in Fig. 10 of Endress and Bischoff indicate the presence of a few atypical observations. Appropriate controls made by Krzanowski's cross-validation program and Q–Q-probability plots failed to pick out atypical values and we conclude that the reduced sets published in the tables accompanying the paper forming the basis of the present analysis do not contain these observations.

The statistical analyses

The main points of interest for our case history pertain to examining the following aspects:

1 Do the individual samples display heterogeneities; i.e., are there distinct groupings visible in the data? This is in response to the indication that several lithologies occur.

2 How well do the compositions agree between chondrites and within chondrites? The conclusions of Endress and Bischoff point to a far-reaching degree of heterogeneity.

Methods

The relevant procedures for the problem at hand are principal coordinate analysis and log-ratio canonical variate analysis. Principal coordinate analysis is useful for disclosing heterogeneities in a sample, and hence, the viability of the lithological groupings in the Ivuna sample. Canonical variate analysis will be used to contrast the data for three meteorites. A constrained discriminant function analysis rounds off the multivariate appraisal of the data for Ivuna, Alais and Orgueil (the published analyses for Tonk are too few for useful statistical appraisal).

The question arises as to the appropriate form of the Q-mode procedure, principal coordinates. There seems to be a general unawareness that even Q-mode methods are affected by the constant-sum constraint. Aitchison (1986) made this amply apparent, but as far as we are aware, his indications have yet to elicit any reactions from the geostatistical consumer community. You can judge matters yourself by analyzing the data in the file *ivunapcd.dat* by the program for constrained principal components and comparing it with the graphs of the results yielded by the same program in its unconstrained version.

The Q-mode analysis – relationships between specimens

The data for the Ivuna meteorite (file *ivunapcd.dat*), were analysed by principal coordinate analysis (using Gower's similarity coefficient) and the program *pcrdcons.exe*. The fit of the plane formed by the first two coordinate axes is good – the residual is 40.31%. The plot of the first two coordinate axes suggests reasonable agreement with the conclusions of Endress and Bischoff in that the points largely group according to the proposed lithological categories (note that data for lithology II are not included because carbonates are completely absent). There is one group encompassing all of lithology IV, one sample of lithology III and one of lithology I. The second grouping contains mainly lithology III, one specimen of lithology I and two of lithology IV. The largest grouping is composed of most of the determinations on lithology I, plus all of the category referred to as "carbonate fragments". This category also contains dolomitic material, but was not classified with any of the lithological units. The present analysis does, however, indicate that on the data available, at least, "carbonate fragments" fall in line with lithology I. The carbonate fragments are said to derive from former carbonate veins, and in contrast to individual carbonate grains, are not considered to be linked to the lithological units. The statistical result probably indicates a chemical though not a genetic agreement and may, therefore, be fortuitous. As an exercise, you can see what the unconstrained version of principal coordinate analysis yields. There is a difference!

Relationships between groups

Constrained canonical variate analysis by means of the log-ratio covariance matrix, using program *cvaconst.exe* and the data in file *chondcva.dat*, provides useful insights into the statistical relationships between meteorites Ivuna, Alais and Orgueil. The log-ratio transformation "consumes" one variable – in the present case, it was considered expedient to let the divisor be the category "CO_2".

A one-way analysis of variance indicates significant differences in means for CaO, MgO, and MnO, but not FeO. This result is reflected in the one-way multivariate analysis of variance, which yields $\chi^2 = 22.9$ for 10 degrees of freedom, which corresponds to $P = 0.01$. The canonical log-contrast discriminant functions are

$z_1 = -7.52\text{CaO} + 13.61\text{Mgo} - 0.12\text{MnO} - 0.12\text{FeO}$
$z_2 = 17.37\text{CaO} + 6.22\text{MgO} + 0.10\text{MnO} + 0.46\text{FeO}$

Obviously, all the discriminatory power lies with CaO and MgO, the dolomite components.

The appearance of the plot of the scores for the first two canonical variates is somewhat unexpected in the light of the published conclusions. We should expect Ivuna and Orgueil to overlap, but this is not so. Orgeuil forms a closely-knit group of its own. Ivuna plots heterogeneously, forming three to four sub-groups. Alais overlaps with part of the points for Ivuna; its values are more closely grouped than those of Ivuna.

Comparison between Ivuna and Alais

The foregoing results motivate a closer examination of the relationship between meteorites Ivuna and Alais. This may be conveniently done by a constrained discriminant function study using the program *dfnconst.exe* and the data in file *chonddfn.dat*. The linear discriminant function is completely dominated by MgO. The likelihood of both samples deriving from the same statistical population is emphatically rejected by $P = 0.016$. However, this is not the whole story, for a direct assessment of likelihood of misidentification indicates that this is fairly high, with $P = 0.3$.

The "internal" test of allocation shows that 25.7% of the Ivuna observations are misidentified and 9.5% of those for Alais. The covariance matrices are not homogeneous ($\chi^2 = 66.35$ for 20 degrees of freedom), which motivates computation of the quadratic discriminant function (note, that the "coefficients" of quadratic discrimination cannot be reified in any meaningful manner). This resulted in an improvement in classificatory performance with 17.14% of the Ivuna observations being wrongly assigned and 4.76% of those for Alais.

Conclusions

As far as the present analysis can be related to the study of Endless and Bischoff (who made no statistical tests), the identification of several lithological categories seems reasonable, though with the suggestion that the carbonate fragments could be reconsidered as being close to lithology I, rather than distinct. Certain reservations remain about the graphical representations and it would be useful if these could be compared and contrasted with those obtainable from the corresponding log-ratio transformed data-matrix. It should be noted that the published lists of analyses permit no more than a gross assessment of the properties of the samples. A full statistical analysis would require access to the chemical determinations on each of the individual components.

The conclusion arrived at by Endless and Bischoff (1996) that there are compositional differences among dolomites between and within CI chondrites is upheld by the statistical study.

CHEMICALLY INDUCED VARIATION IN THE SANTONIAN OSTRACOD *VEENIA FAWWARENSIS*

Introduction

This example is an illustration of how palaeontology and geochemistry can be united to yield a useful palaeoenvironmental analysis. Ostracods are well known to exhibit marked variability in features of the carapace, some of which are ecophenotypic, that is, the outcome of the influence of chemical factors on the morphology of the shell. The data analysed by Abe et al. (1988) consists of measurements made on the carapace of *Veenia fawwarensis* HONIGSTEIN from nine levels in a quarry at Shiloah, Jerusalem. The carapaces show an interesting variation in shape, which could be largely related to the content of magnesium in the limestone. The three levels selected here for the first part of the study come from the initial part of the sequence in which the chemical conditions seem to have been more stable than in younger beds. The aim of the analysis is to see whether this presumed stability is reflected in the variability of the carapaces. If the hypothesis is correct then the data should display evidence of homogeneity for that part of the sequence in which the magnesium content is low, but significant heterogeneity where the content of Mg exceeds the normal level. The method of common principal component analysis offers a unique means of not only checking the data for homogeneity but also for obtaining information on covariance relationships.

The measurements made on the carapaces are: (1) the length of the carapace; (2) the height of the carapace; (3) the distance from the eye-tubercle to the adductorial boss; (4) the distance from the adductorial boss to the posteroventral angle; (5) the distance from the posterodorsal angle to the posteroventral angle; (6) the distance from the eye tubercle to the posterodorsal angle; and (7) the maximum breadth of the carapace.

The full data-set is in the file **israelpc.dat** from which the appropriate combinations of matrices were selected by hand after processing of the observations by means of **covpc.exe** which yielded the three covariance matrices listed below. They were analysed by means of the program for common principal component analysis, **cpca.exe.**

Findings

Common principal component analysis for **Veenia fawwarensis**

The first three levels

Groups = 3; dimensions = 7; $N = 79$.

The input consists of three covariance matrices. To illustrate the comparisons following on in the analysis, one of these is listed below.

Usual principal component results for first sample

Standard latent roots

| 223.41739 | 66.40113 | 31.51319 | 14.44023 | 7.62932 | 4.70702 | 4.02932 |

Standard latent vectors

	1	2	3	4	5	6	7
1	0.4435	−0.1206	−0.1337	0.0650	−0.6841	0.4427	−0.3204
2	0.2061	0.0151	0.2762	−0.3182	0.5367	0.6962	−0.0840
3	0.1946	0.9252	0.2314	−0.0159	−0.1711	−0.0269	0.1492
4	0.3993	−0.2473	0.0165	−0.0665	−0.0969	0.0304	0.8743
5	0.4952	−0.1064	0.1192	−0.6019	0.0447	−0.5349	−0.2807
6	0.5016	0.1418	−0.5802	0.4225	0.4445	−0.0822	−0.0938
7	0.2581	−0.1913	0.7079	0.5908	0.0763	−0.1572	−0.1265

Usual principal component results for second sample

Standard latent roots

| 272.42235 | 46.43150 | 19.26334 | 9.65157 | 3.98853 | 1.94177 | 1.23534 |

Standard latent vectors

	1	2	3	4	5	6	7
1	0.4518	0.2892	−0.0467	0.1044	0.3246	−0.7388	0.2189
2	0.2162	−0.0848	0.1759	−0.4275	−0.5072	0.0088	0.6893
3	0.3704	−0.8474	−0.2254	−0.1832	0.1420	−0.1049	−0.1707
4	0.3672	0.2746	0.1241	−0.3480	−0.4516	−0.1188	−0.6596
5	0.4143	0.2671	0.0104	−0.3709	0.5455	0.5636	0.0645
6	0.4993	0.0737	−0.3797	0.6157	−0.3330	0.3232	0.0820
7	0.2371	−0.1972	0.8697	0.3675	0.0693	0.0826	−0.0427

The input matrix No. 3

	1	2	3	4	5	6	7
1	57.8367	12.7128	29.9954	48.6790	38.9729	69.4328	8.9896
2	12.7128	11.1034	27.5848	9.4820	12.9943	18.0854	8.6065
3	29.9954	27.5848	105.8183	12.1167	31.6062	48.6308	8.6924
4	48.6790	9.4820	12.1167	47.9749	33.3990	56.9120	9.3373
5	38.9729	12.9943	31.6062	33.3990	40.0647	41.3932	14.0254
6	69.4328	18.0854	48.6308	56.9120	41.3932	96.5092	7.9927
7	8.9896	8.6065	8.6924	9.3373	14.0254	7.9927	28.2548

Usual principal component results

Standard latent roots

| 250.97120 | 83.55588 | 32.59253 | 13.59158 | 3.63751 | 1.92377 | 1.28952 |

Standard latent vectors

	1	2	3	4	5	6	7
1	0.4494	0.2595	−0.0447	−0.1133	0.2364	−0.6107	0.5358
2	0.1528	−0.1696	0.1636	0.1039	−0.4731	0.4412	0.7013
3	0.4262	−0.8459	−0.0860	−0.0518	−0.0857	−0.2258	−0.1855
4	0.3633	0.3822	0.0758	−0.1507	−0.7273	−0.1574	−0.3738
5	0.3347	0.0469	0.3884	−0.6709	0.3285	0.4108	−0.0897
6	0.5827	0.1998	−0.3546	0.4919	0.2495	0.4074	−0.1567
7	0.0979	−0.0119	0.8255	0.5088	0.1171	−0.1479	−0.1202

Correlation matrices of common principal components

Sample No. 1

1	1.00000	−0.08524	0.15852	−0.02433	0.11389	0.09944	−0.20749
2	−0.08524	1.00000	0.28976	0.16055	0.00680	−0.14289	−0.07784
3	0.15852	0.28976	1.00000	0.18079	0.03848	−0.08825	0.13203
4	−0.02433	0.16055	0.18079	1.00000	−0.14522	0.01301	0.21061
5	0.11389	0.00680	0.03848	−0.14522	1.00000	0.11214	−0.09656
6	0.09944	−0.14289	−0.08825	0.01301	0.11214	1.00000	−0.05341
7	−0.20749	−0.07784	0.13203	0.21061	−0.09656	−0.05341	1.00000

Sample No. 2

1	1.00000	−0.11910	−0.04563	0.14504	0.01360	0.21191	−0.21717
2	−0.11910	1.00000	0.02558	0.14932	0.10007	−0.11506	−0.28638
3	−0.04563	0.02558	1.00000	−0.07482	−0.19059	−0.00135	0.10573
4	0.14504	0.14932	−0.07482	1.00000	−0.28991	−0.29437	−0.15001
5	0.01360	0.10007	−0.19059	−0.28991	1.00000	0.12816	0.01936
6	0.21191	−0.11506	−0.00135	−0.29437	0.12816	1.00000	0.27175
7	−0.21717	−0.28638	0.10573	−0.15001	0.01936	0.27175	1.00000

Sample No. 3

1	1.00000	0.07112	−0.13436	−0.06484	−0.03250	−0.23092	0.33792
2	0.07112	1.00000	−0.20491	−0.19617	−0.15519	0.13943	0.23449
3	−0.13436	−0.20491	1.00000	−0.12073	0.18646	0.10910	−0.34712
4	−0.06484	−0.19617	−0.12073	1.00000	0.32423	0.21561	−0.07686
5	−0.03250	−0.15519	0.18646	0.32423	1.00000	−0.18435	0.05220
6	−0.23092	0.13943	0.10910	0.21561	−0.18435	1.00000	−0.09900
7	0.33792	0.23449	−0.34712	−0.07686	0.05220	−0.09900	1.00000

All three correlation matrices contain small to relatively small entries thus evidencing to the satisfactory fit of the common principal component model to the data.

Sample   The CPCA latent roots, sample by sample

1	5.10176	8.79927	218.44039	67.16904	4.70165	14.10178	33.82372
2	5.22716	2.20728	271.81620	44.38965	1.61215	10.52390	19.15807
3	4.36570	2.33852	243.65097	82.80153	1.84278	14.15826	38.40424

The log-likelihood test for a valid CPC model
Chi-square = 49.1976 for df = 42 (the 5% significance for computed chi-square is 57.8397).

This result shows that the common principal component model is not contravened. This is, however, not supported by the following test for equal covariance matrices.

Test for equality of covariance matrices

Covariance matrices chi-square = 31.2451 for df = 49. The 5% significance for computed chi-square is 66.0544.

This result is not truly indicative of a CPC model. We shall however, continue the illustration of the calculations.

Ordered latent roots for common principal components

218.44039	67.16904	33.82372	14.10178	8.79927	5.10176	4.70165
271.81620	44.38965	19.15807	10.52390	5.22716	2.20728	1.61215
243.65097	82.80153	38.40424	14.15826	4.36570	2.33852	1.84278

Common latent vectors

	1	2	3	4	5	6	7
1	0.4593	0.2293	−0.1338	−0.0463	−0.6000	0.1902	−0.5659
2	0.1921	−0.1034	0.2362	−0.1181	0.4601	−0.5518	−0.6054
3	0.3408	−0.8967	−0.0130	−0.0957	−0.2197	−0.0394	0.1438
4	0.3822	0.3273	0.0213	−0.1145	−0.2528	−0.6502	0.4965
5	0.4191	0.1481	0.2200	−0.6366	0.3482	0.4419	0.1796
6	0.5175	0.0492	−0.4878	0.5286	0.4371	0.1170	0.0879
7	0.2159	0.0349	0.7996	0.5262	−0.0623	0.1613	0.0774

Standard errors of common principal component coefficients

	1	2	3	4	5	6	7
1	0.0219	0.0469	0.0569	0.0911	0.0765	0.3006	0.1202
2	0.0181	0.0456	0.0466	0.0864	0.1175	0.3134	0.2828
3	0.0667	0.0275	0.1316	0.0653	0.0402	0.0840	0.0392
4	0.0267	0.0366	0.0623	0.0675	0.1229	0.2530	0.3282
5	0.0241	0.0625	0.0911	0.0650	0.1142	0.1157	0.2291
6	0.0268	0.0885	0.0769	0.0874	0.0798	0.0934	0.0866
7	0.0373	0.1213	0.0691	0.1044	0.0894	0.0627	0.0912

Reconstituted covariance matrix No. 1

	1	2	3	4	5	6	7
1	55.1022	15.3312	21.2406	42.7439	41.8546	52.1056	18.5114
2	15.3312	15.9984	19.3925	13.5159	19.0256	17.7813	13.3984
3	21.2406	19.3925	80.0427	9.8376	22.4014	34.2523	13.0468
4	42.7439	13.5159	9.8376	43.1804	37.6041	41.9201	18.2995
5	41.8546	19.0256	22.4014	37.6041	49.3987	41.1586	21.5714
6	52.1056	17.7813	34.2523	41.9201	41.1586	72.4272	15.1353
7	18.5114	13.3984	13.0468	18.2995	21.5714	15.1353	35.9879

Corresponding matrix of residuals

	1	2	3	4	5	6	7
1	−4.6508	2.2633	−9.7134	−1.8840	6.0164	−3.0605	5.6100
2	2.2633	1.8767	−8.2548	4.4755	5.9928	0.0456	1.5816
3	−9.7134	−8.2548	−12.7295	−6.8848	−6.5572	−8.6872	−8.7025
4	−1.8840	4.4755	−6.8848	−0.2666	7.8620	−0.8746	7.1479
5	6.0164	5.9928	−6.5572	7.8620	13.4966	7.9492	6.4286
6	−3.0605	0.0456	−8.6872	−0.8746	7.9492	−0.1148	3.0238
7	5.6100	1.5816	−8.7025	7.1479	6.4286	3.0238	2.3885

Reconstituted covariance matrix No. 2

	1	2	3	4	5	6	7
1	61.5350	21.7713	33.6194	50.2785	53.3818	65.5491	25.1704
2	21.7713	14.3663	21.7175	19.8228	21.8885	23.9453	13.4693
3	33.6194	21.7175	67.5101	22.8572	33.2949	45.3516	17.8941
4	50.2785	19.8228	22.8572	47.3524	44.9876	53.0617	22.1731
5	53.3818	21.8885	33.2949	44.9876	55.2382	54.2988	25.0101
6	65.5491	23.9453	45.3516	53.0617	54.2988	80.8934	25.9442
7	25.1704	13.4693	17.8941	22.1731	25.0101	25.9442	28.0389

Corresponding matrix of residuals

	1	2	3	4	5	6	7
1	−0.3548	2.6256	0.9025	−2.4490	0.7248	−3.0179	0.9132
2	2.6256	2.6707	2.9915	2.9266	1.9814	2.0888	2.5252
3	0.9025	2.9915	4.6506	3.3749	−1.2126	2.4156	9.3950
4	−2.4490	2.9266	3.3749	−4.2726	−0.0358	−4.6953	−0.2339
5	0.7248	1.9814	−1.2126	−0.0358	−2.0372	0.3142	−1.6018
6	−3.0179	2.0888	2.4156	−4.6953	0.3142	−5.6464	1.4066
7	0.9132	2.5252	9.3950	−0.2339	−1.6018	1.4066	4.9898

Reconstituted covariance matrix No. 3. This is to be checked against the third input matrix, listed above.

	1	2	3	4	5	6	7
1	58.0563	17.9211	21.3672	48.2442	48.6826	60.3911	20.5058
2	17.9211	14.7120	23.3642	16.2079	20.5129	18.5765	15.6334
3	21.3672	23.3642	95.1691	7.9507	24.3502	38.6227	14.2460
4	48.2442	16.2079	7.9507	47.1116	42.9501	47.7515	20.4983
5	48.6826	20.5129	24.3502	42.9501	53.3938	45.1621	24.7676
6	60.3911	18.5765	38.6227	47.7515	45.1621	79.0544	16.3501
7	20.5058	15.6334	14.2460	20.4983	24.7676	16.3501	40.0648

Corresponding matrix of residuals

	1	2	3	4	5	6	7
1	−0.2196	−5.2083	8.6282	0.4348	−9.7097	9.0417	−11.5162
2	−5.2083	−3.6086	4.2206	−6.7259	−7.5186	−0.4911	−7.0269
3	8.6282	4.2206	10.6492	4.1660	7.2560	10.0081	−5.5536
4	0.4348	−6.7259	4.1660	0.8633	−9.5511	9.1605	−11.1610
5	−9.7097	−7.5186	7.2560	−9.5511	−13.3291	−3.7689	−10.7422
6	9.0417	−0.4911	10.0081	9.1605	−3.7689	17.4548	−8.3574
7	−11.5162	−7.0269	−5.5536	−11.1610	−10.7422	−8.3574	−11.8100

The effect of Mg on the shell becomes apparent when the samples from all nine levels are considered. The following results extracted from that analysis indicate that to be so. Sampling level 7 has the highest content of Mg being two and a half times that of the adjacent sampling levels.

Correlation matrices of common principal components

Sample No. 4

	1	2	3	4	5	6	7
1	1.0000	0.5088	0.0840	−0.4225	−0.2013	0.1198	0.0575
2	0.5088	1.0000	0.1001	−0.0760	0.1063	−0.1440	−0.2064
3	0.0840	0.1001	1.0000	−0.3227	−0.1317	−0.2591	0.0598
4	−0.4225	−0.0760	−0.3227	1.0000	0.3968	0.0586	−0.5327
5	−0.2013	0.1063	−0.1317	0.3968	1.0000	0.1025	−0.3324
6	0.1198	−0.1440	−0.2591	0.0586	0.1025	1.0000	0.1893
7	0.0575	−0.2064	0.0598	−0.5327	−0.3324	0.1893	1.0000

There are some large residual correlations here and also in the residual correlations for sample 7.

Sample No. 7

	1	2	3	4	5	6	7
1	1.0000	−0.3377	−0.0625	0.2122	−0.0254	−0.4871	0.0178
2	−0.3377	1.0000	−0.1290	−0.5105	0.1692	0.1145	0.0327
3	−0.0625	−0.1290	1.0000	−0.0367	0.1979	0.0144	0.1008
4	0.2122	−0.5105	−0.0367	1.0000	0.0727	−0.0835	−0.2228
5	−0.0254	0.1692	0.1979	0.0727	1.0000	−0.0540	−0.0661
6	−0.4871	0.1145	0.0144	−0.0835	−0.0540	1.0000	0.2529
7	0.0178	0.0327	0.1008	−0.2228	−0.0661	0.2529	1.0000

The log-likelihood test for adequacy of the common principal component model yields a value of chi-square = 230.58 for 168 degrees of freedom which, when compared with the 5% significance level of 198.96, is significant and therefore does not support the common principal component model. The test for equality of covariance matrices gives a value of chi-square = 136.54, which for 49 degrees of freedom exceeds by far the 5% significance level of 66.05. This result does not contravene the CPC model.

Conclusions

The common principal component analysis of the ostracod data supports an hypothesis of ecophenotypic response in the shell morphology in relation to the content of magnesium in the original seawater. The CPC model is hardly appropriate for analysing the low-Mg levels, but more adequate for the levels in which the content of Mg is relatively elevated. Although one can hardly be adamant about the reason for this, without more extensive sampling, one possible suggestion that arises is that the effect of excess magnesium in the environment has been to exaggerate variability in the carapace.

MORPHOMETRICAL RELATIONSHIPS BETWEEN SCAPHITID MACROCONCHS AND MICROCONCHS

Introduction

One of the best known heteromorphic ammonite groups are the Scaphitidae. The scaphitids display sexual dimorphism of the type commonly interpreted as such in ammonites. Dimorphs are referred to as **macroconchs** and **microconchs**, the (arbitrary) convention being that the macroconchs are said to be females and the microconchs males. The dimorphs are usually claimed to be distinguishable at maturity by differences in the shape of the body chamber. In macroconchs, the body chamber increases gradually in width and abruptly in height. In microconchs, the body chamber increases gradually in size through the shaft and then decreases slightly towards the aperture. In average, adult macroconchs are larger than adult microconchs.

Well preserved scaphitids occur in the Upper Cretaceous (Maastrichtian) Fox Hills Formation in South Dakota, U.S.A. Landman and Waage (1993) studied the ontogeny of several species. In the present case history, we report on those referred to the genera *Discoscaphites* and *Hoploscaphites*.

Our analysis is concerned with searching for multidimensional differentiation between microconchs and macroconchs of the same species and to compare and contrast eventual differences between species of the same genus. Landman and Waage concluded that macroconchs tend to be more sharply differentiated between species than are microconchs. We shall also test that hypothesis.

The data

Landman and Waage supplied their monograph with many tables of measurements on the dimensions commonly observed on ammonites, to wit: shell diameter, whorl-width, whorl-height and diameter of the umbilicus. From Appendix II (Landman and Waage, 1993, p. 248) we have extracted measurements on ontogenetic series for the following species:

Hoploscaphites nicolletti (Morton)
H. comprimus (Owen)
Discoscaphites conradi (Morton)
D. rossi (Landman and Waage)
D. gulosus (Morton)

Findings

The data-sets are in the form of measurements made through the ontogeny of adult dimorphs. Hence each 'sample' is made up of four variables observed over the coiling history of the specimen. These quadrivariate series form the building blocks of our analysis.

The statistical method

The method of canonical variate analysis is the appropriate procedure for testing our hypotheses. The program **cva.exe** produces, as subsidiary computations, principal component analyses, a multivariate analysis of variance and the full set of Mahalanobis distances and associated significances. The graphical analysis centres around the plot of the canonical variate means and the superimposed minimum spanning tree for indicating nearest neighbour relationships. It is often necessary to remind oneself that a two-dimensional plot of multivariate scores can yield a spurious picture of true relationships between points. Think of stars in the sky. Two stars that seem to be near each other to the eye in a two-dimensional frame may actually lie very widely apart in the third dimension.

FINDINGS

The first data-set contains the observations on two species of *Discoscaphites*. The data are stored in the file **Discosc.dat.**

The aim of the analysis is to ascertain whether macroconchs and microconchs differ, on average, with respect to the four standard distance-measures. The appropriate method of analysis is to apply canonical variate analysis to the logarithmically transformed observations; the transformation tends to reduce the ontogenetic size-differences. There are nine samples, each comprising a suite of observations on stages in the ontogeny of the ammonite shell. It is instructive to begin with an examination of the principal components of a macroconch and a microconch selected at random from the groups.

Summary for principal components of macroconchs for *D. conradi*

Latent Roots
1.0586   0.0125   0.0005   0.0002

Percentage for root 1 = 98.76
Percentage for root 2 = 1.17
Percentage for root 3 = 0.05
Percentage for root 4 = 0.02

The first latent root accounts for almost 98% of the total variability in the observations. The third and fourth latent roots are almost zero. This result indicates that the covariance matrix is very close to being of rank 2, which is a normal condition in cephalopod shells. It is a result of the rigid regime imposed by the logarithmic growth spiral. Turning now to the first latent vector, we see that there appears to be an allometric relationship between the umbilical diameter and the three other variables. There is some further information to be extracted from the latent vectors. The fourth latent vector is connected to an almost zero root and it can therefore be interpreted as representing an invariant linear combination of the variables in which the umbilical diameter plays no significant part.

Principal component loadings

	1	2	3	4
1	0.5376	−0.1055	−0.5623	−0.6195
2	0.4390	−0.1953	0.8145	−0.3251
3	0.6216	−0.3108	−0.1271	0.7077
4	0.3632	0.9242	0.0652	0.0986

The microconchs yield a somewhat different result. The latent roots are, as only to be expected, similar:

Latent roots for microconchs of *D. conradi*

1.231295   0.014642   0.001600   0.000282

Percentage for 1 = 98.68
Percentage for 2 = 1.17
Percentage for 3 = 0.13
Percentage for 4 = 0.02

Principal component loadings

	1	2	3	4
1	0.5379	−0.1269	−0.1123	−0.8258
2	0.4109	−0.0532	−0.8233	0.3879
3	0.6017	−0.4660	0.5158	0.3934
4	0.4239	0.8740	0.2085	0.1135

The elements of the first latent vector are roughly equal which does not suggest the presence of allometric relationships. The pattern for the invariant vector also differs from that of macroconchs. These results are echoed by the other samples.

Findings

Sample sizes and corresponding mean vectors

1	10.	0.524	0.217	0.149	−0.059
2	11.	0.635	0.306	0.266	0.026
3	10.	0.627	0.278	0.253	0.047
4	9.	0.601	0.283	0.242	−0.032
5	10.	0.617	0.257	0.271	0.009
6	8.	0.465	0.144	0.118	−0.155
7	8.	0.461	0.144	0.099	−0.098
8	12.	0.763	0.483	0.398	0.170
9	12.	0.742	0.477	0.350	0.206

A univariate one-way analysis of variance yields the result that samples are homogeneous with respect to means for all species and dimorphs considered in the same connection.

Output for canonical variate latent roots and vectors

Canonical Root	1	1.0939	percentage	90.71
Canonical Root	2	0.0642	percentage	5.32
Canonical Root	3	0.0386	percentage	3.20
Canonical Root	4	0.0092	percentage	0.76

As to be expected, the first canonical root dominates greatly over the other three. The fourth root is close to zero.

Canonical Vectors

	1	2	3	4
1	1.1633	−3.4302	−2.6118	−0.2912
2	−2.2359	0.9657	−0.2428	−0.1278
3	0.6819	2.2671	1.7398	0.5265
4	−0.1143	0.1187	1.1721	−0.0028

The greatest loadings in the first canonical vector are for length and height of the shell. The rôle of the umbilicus is insignificant.

Variance-weighted canonical variate coefficients

	1	2	3	4
1	10.4698	−30.8715	−23.5061	−2.6207
2	−20.1232	8.6911	−2.1853	−1.1502
3	6.1373	20.4039	15.6580	4.7382
4	−1.0289	1.0687	10.5485	−0.0252

D-square above diagonal, D below diagonal

	1	2	3	4	5	6	7	8	9
1	0.0000	0.0610	0.3808	0.0872	0.9357	0.7859	0.5509	2.5972	4.0455
2	0.2470	0.0000	0.2795	0.1106	0.9791	0.9811	0.7680	2.6755	4.0695
3	0.6171	0.5287	0.0000	0.5957	0.4858	0.9203	0.5699	4.2795	5.7518
4	0.2953	0.3325	0.7718	0.0000	1.0330	0.7143	0.7306	2.6340	4.2980
5	0.9673	0.9895	0.6970	1.0164	0.0000	0.3057	0.2302	5.9813	8.1953
6	0.8865	0.9905	0.9593	0.8451	0.5529	0.0000	0.2573	5.4967	7.9321
7	0.7422	0.8764	0.7549	0.8548	0.4798	0.5072	0.0000	4.6082	6.5915
8	1.6116	1.6357	2.0687	1.6230	2.4457	2.3445	2.1467	0.0000	0.3528
9	2.0113	2.0173	2.3983	2.0732	2.8627	2.8164	2.5674	0.5940	0.0000

Minimum spanning tree

OTU	connected to OTU	distance
2	1	0.06102
4	1	0.08720
3	2	0.27948
5	3	0.48585
7	5	0.23018
6	7	0.25730
8	1	2.59718
9	8	0.35279

(OTU is an acronym for Operational Taxonomic Unit, a term borrowed from numerical taxonomy.)

Hotellings T^2 test of significance

F-ratio for D-square and associated probabilities

	1	2	3	4	5	6	7	8	9
1	0.0000	0.0769	0.4583	0.0994	1.1263	0.8409	0.5894	3.4104	5.3123
2	0.9860	0.0000	0.3524	0.1318	1.2347	1.0939	0.8563	3.6967	5.6227
3	0.7685	0.8427	0.0000	0.6793	0.5848	0.9846	0.6098	5.6195	7.5528
4	0.9791	0.9672	0.6113	0.0000	1.1780	0.7283	0.7450	3.2611	5.3214
5	0.3506	0.3030	0.6776	0.3271	0.0000	0.3271	0.2463	7.8542	10.7615
6	0.5055	0.3661	0.5779	0.5780	0.8597	0.0000	0.2478	6.3517	9.1660
7	0.6743	0.5040	0.6598	0.5669	0.9101	0.9093	0.0000	5.3250	7.6168
8	0.0128	0.0085	0.0006	0.0160	0.0000	0.0002	0.0009	0.0000	0.5096
9	0.0009	0.0006	0.0000	0.0009	0.0000	0.0000	0.0000	0.7316	0.0000

The above array shows that on average macroconchs and microconchs of *D. conradi* and *D. rossi* do not differ with respect to the characters measured on them. However, the macroconchs and microconchs of both of these species differ highly significantly from macroconchs and microconchs of *D. gulosus*. Another interesting result is that macroconchs and microconchs of all three species do not differ significantly from each other. Interesting information is available in the *Graph Server* plot of the generated file *canmeans*. Superimposition of the minimum spanning tree on the plot for the first two transformed mean vectors locates one of the microconch individuals for *conradi* and the macroconch and microconchs for *rossi* on the same branch. The *conradi* macroconchs join to *conradi* microconchs, although the latter are separated. The *gulosus* means are located on a separate branch.

Significance of canonical roots

Root No 1 = 0.8483
Chi-square = 69.98 for 32 degrees of freedom.
Probability = 0.0001

Root No 2 = 0.1093
Chi-square = 9.01 for 21 degrees of freedom.
Probability = 0.989

The test for significant roots indicates that there is only one that is statistically different from nought. Hence, practically all of the discrimination between groups lies with the first canonical variate.

MANOVA test of equality of means

Wilk's Lambda = 0.428
Chi-square = 67.448 with degrees of freedom = 32; probability = 0.0003.

The result of this test, which compares all centroids against each other, is that they are highly significantly different. A standard test for homogeneity of covariance matrices yields a value of chi-square of 147.93, which for 80 degrees of freedom is highly significant, thus pointing to pronounced heterogeneity in the covariance matrices.

We shall now examine the material of *Hoploscaphites*. The data are in the file **hoplo.dat**. A selection taken from the suite of principal component analyses is included below to demonstrate that there is a slight difference in relation to the data for *Discoscaphites*. This resides mainly in the approximately isometric growth relationship in the first (and completely dominating) principal component.

H. nicolletti macroconchs 1

Summary for principal components

Latent roots
1.20713 0.02496 0.00011 0.00002

Percentage for 1 = 97.96
Percentage for 2 = 2.03
Percentage for 3 = 0.0085
Percentage for 4 = 0.0016

Principal component loadings

	1	2	3	4
1	0.5283	−0.1517	−0.4447	−0.7072
2	0.4374	−0.2563	−0.5045	0.6989
3	0.5603	−0.3821	0.7339	0.0390
4	0.4644	0.8748	0.0956	0.0991

The almost invariant, fourth latent vector expresses a fixed relationship between length and height of the shell.

H. nicolletti macroconch 2

Summary for principal components

Latent roots

0.95368   0.00350   0.00018   0.00005

Percentage for 1 = 99.61
Percentage for 2 = 0.37
Percentage for 3 = 0.018
Percentage for 4 = 0.005

Principal component loadings

	1	2	3	4
1	0.5274	−0.1434	−0.2023	−0.8126
2	0.4356	−0.2261	−0.7134	0.5002
3	0.5728	−0.3906	0.6662	0.2747
4	0.4517	0.8808	0.0794	0.1180

H. nicolletti microconch 1

Summary for principal components

Latent roots

1.04956   0.00680   0.00054   0.000005

Findings

Percentage for 1 = 99.30
Percentage for 2 = 0.64
Percentage for 3 = 0.05
Percentage for 4 = 0.0005

Principal component loadings

	1	2	3	4
1	0.5220	−0.0842	−0.2368	−0.8151
2	0.4282	−0.3859	0.8135	0.0778
3	0.5636	−0.3470	−0.5135	0.5460
4	0.4759	0.8506	0.1361	0.1773

H. nicolletti microconch 2

Summary for principal components

Latent roots

1.04305   0.00590   0.00029   0.00003

Percentage for 1 = 99.4069
Percentage for 2 = 0.5625
Percentage for 3 = 0.0277
Percentage for 4 = 0.0030

Principal component loadings

	1	2	3	4
1	0.5197	−0.1062	−0.2799	−0.8002
2	0.4204	−0.4373	0.7932	0.0536
3	0.5579	−0.3099	−0.5058	0.5804
4	0.4918	0.8376	0.1915	0.1413

A general observation for *Hoploscaphites* is that the covariance matrices are very nearly of rank 1. Moreover, the covariation pattern is to all intents and purposes the same for macroconchs as for microconchs.

Sample sizes and corresponding mean vectors

1	10.	0.719	0.379	0.374	0.088
2	9.	0.672	0.332	0.311	0.110
3	10.	0.685	0.351	0.320	0.103
4	10.	0.692	0.347	0.326	0.113
5	10.	0.547	0.180	0.155	0.042

6	10.	0.511	0.154	0.117	0.007
7	8.	0.784	0.385	0.443	0.180
8	9.	0.727	0.352	0.391	0.107
9	10.	0.672	0.264	0.268	0.197
10	10.	0.588	0.216	0.209	0.036

Just as for *Discoscaphites*, the univariate analysis of variance does not disclose any significant differences in means.

Output for CVA latent roots and vectors

Canonical Root	1	1.3682 percentage	76.47
Canonical Root	2	0.3834 percentage	21.43
Canonical Root	3	0.0220 percentage	1.23
Canonical Root	4	0.0153 percentage	0.86

We have however here a rather surprising result in that the first canonical root is relatively much smaller, and the second canonical root is relatively very much greater, than in the case of the material for *Discoscaphites*. Another property of the present material is that the first and second canonical vectors reflect covariation in length, height and breadth of the test. As before, umbilical diameter is not important.

Canonical vectors

	1	2	3	4
1	−6.7326	−5.3306	−4.8843	0.1915
2	4.2987	−1.9317	0.5882	0.0100
3	2.3941	5.8711	2.8666	−0.1465
4	0.6854	0.6944	1.4551	0.1771

Variance-weighted canonical variate coefficients

	1	2	3	4
1	−62.4375	−49.4328	−45.2943	1.7762
2	39.8657	−17.9130	5.4550	0.0929
3	22.2027	54.4446	26.5837	−1.3583
4	6.3568	6.4395	13.4939	1.6426

D-square above diagonal, D below diagonal

	1	2	3	4	5	6	7	8	9	10
1	0.0000	0.2964	0.5573	0.6237	5.5949	5.2175	4.0590	2.5236	11.4563	5.4548
2	0.5444	0.0000	0.4658	0.4209	4.5103	4.1830	3.7041	2.4688	9.8496	4.7438
3	0.7466	0.6825	0.0000	0.1733	5.2232	4.7552	5.5472	4.3076	10.0121	5.4148
4	0.7898	0.6488	0.4163	0.0000	3.5331	3.1619	3.9783	3.1768	7.7224	3.6627

Findings 263

5	2.3654	2.1238	2.2854	1.8796	0.0000	0.0220	1.8154	3.0204	1.3469	0.1448
6	2.2842	2.0452	2.1806	1.7782	0.1484	0.0000	1.9630	3.0557	1.4822	0.1736
7	2.0147	1.9246	2.3552	1.9946	1.3474	1.4011	0.0000	0.3754	5.4171	1.4603
8	1.5886	1.5712	2.0755	1.7823	1.7379	1.7481	0.6127	0.0000	7.9417	2.6276
9	3.3847	3.1384	3.1642	2.7789	1.1606	1.2175	2.3275	2.8181	0.0000	1.6308
10	2.3355	2.1780	2.3270	1.9138	0.3805	0.4166	1.2084	1.6210	1.2770	0.0000

Minimum spanning tree

OTU connected to OTU distance

2	1	0.29637
4	2	0.42093
3	4	0.17334
8	2	2.46877
7	8	0.37543
10	7	1.46026
5	10	0.14480
6	5	0.02201
9	5	1.34694

Length of spanning tree = 6.78

Hotellings test of significance

F-ratio for *D*-square and associated Probabilities

	1	2	3	4	5	6	7	8	9	10
1	0.0000	0.3387	0.6724	0.7525	6.7497	6.2944	4.3527	2.8843	13.8208	6.5806
2	0.8520	0.0000	0.5323	0.4811	5.1549	4.7808	3.7852	2.6805	11.2571	5.4217
3	0.6159	0.7152	0.0000	0.2091	6.3012	5.7366	5.9485	4.9232	12.0786	6.5324
4	0.5618	0.7522	0.9309	0.0000	4.2623	3.8145	4.2661	3.6307	9.3162	4.4186
5	0.0001	0.0010	0.0002	0.0036	0.0000	0.0266	1.9467	3.4520	1.6249	0.1747
6	0.0002	0.0017	0.0005	0.0069	0.9969	0.0000	2.1050	3.4924	1.7881	0.2094
7	0.0032	0.0073	0.0003	0.0036	0.1094	0.0867	0.0000	0.3837	5.8090	1.5659
8	0.0273	0.0370	0.0014	0.0091	0.0118	0.0111	0.8214	0.0000	9.0765	3.0031
9	0.0000	0.0000	0.0000	0.0000	0.1747	0.1380	0.0004	0.0000	0.0000	1.9674
10	0.0001	0.0007	0.0002	0.0029	0.9484	0.9308	0.1901	0.0229	0.1062	0.0000

The comparisons between the macroconchs of *H. nicolletti* all give non-significant variance ratios, and is a greater number of significantly different comparisons than for the first species. Examine the plot of the canonical means and draw in the minimum spanning network as an exercise. The superimposition of the minimum spanning tree on the plot for the first two canonical variate means puts all four macroconchs of *nicolletti* on the same branch, which joins to the branch with the two specimens of macroconch *comprimus*. The microconch specimens for both species are located on the same branch.

Significance of latent roots

Root No 1 = 1.2238
Chi-square = 107.69 for 36 degrees of freedom
Probability < 0.0001

Root No 2 = 0.3617
Chi-square = 31.83 for 24 degrees of freedom.
Probability = 0.131

There is one clearly significant canonical root (as was also found for the other genus) and one that is not significant, but which is connected to a fairly large value of chi-squared. Harris (1975, p. 108) has pointed out that the "test for subsequent roots" in canonical variate analysis is conservative.

MANOVA test of equality of means

Wilk's Lambda = 0.294

Chi-square = 104.02 with degrees of freedom = 36 and $P < 0.00001$. Hence, the ten centroids differ significantly.

The test of homogeneity of covariance matrices yields a chi-square of 184.93, which for 90 degrees of freedom, is highly significant; $P < 0.00001$.

CONCLUSIONS FOR THE SCAPHITIDS

The foregoing analysis shows that there are subtle differences between macroconchs and microconchs (for the dimensions available for study), but these are not as extreme as would seem to be inferred by the descriptive account of Landman and Waage (1993). The multivariate analysis succeeded in accessing information on variational patterns in the macroconchs and microconchs which seem to be species-specific. Perhaps the most valuable information derives from the minimum spanning tree in relation to the plot of the first two canonical variate means. These graphs express clearly the discrimination between macroconchs and microconchs.

A 225 KYR RECORD OF DUST SUPPLY FOR THE NORTHERN ARABIAN SEA

Introduction

Reichart et al. (1997) have been concerned with interpreting the history of monsoonal circulation in the Indian Ocean. Previous work in the western Arabian Sea had shown that the analysis of proxy records for the productivity of surface waters and the input of dust revealed that past variations in the summer insolation of the northern hemisphere, changes in climatic boundary conditions and glacial cycles caused variations in monsoonal intensity. Sediments from the northern edge of the Arabian Sea, collected by the Netherlands Ocean Programme in 1992 from the Murray Ridge were thought to be a useful adjunct in that they derive from open-sea conditions and out of the range of turbidity. The data studied by the Dutch group were obtained from a 15 m long sediment core which encompasses more than two complete glacial cycles. The dust originates from the bounding deserts, being transported by the Shamal winds that blow from the Arabian Peninsula at times of the year.

The chemical material

The sedimentary sequence consists largely of a succession of alternating dark green to light green hemipelagic muds. The weighed samples were dried and then determined with respect to the elements Al, Ba, Be, Ca, Fe, K, Mg, Mn, P, S, Sr, Ti, Y, Zn and Zr. The chemical data, published in Appendix A of the article under review, were analysed statistically by principal component analysis with varimax rotation of the latent vectors (we are not told this by the authors, but a control of the published information shows that this must have been the case in their study). The principal components were extracted from the raw correlations, notwithstanding the fact that the data are compositional in nature. Inasmuch as the palaeoclimatological interpretation of the sedimentary relationships is based on the reification of the results of the rotated principal component analysis, it can be of interest to compare and contrast the original conclusions with the findings yielded by an appropriately formulated analytical procedure. We point out that in spite of the data being quoted in percent and ppm, they can be easily treated in the same connection when adjusted to fit the same scale, granted that they were obtained in the same suite of analyses and are therefore not an example of joint variability, such as occurs when the determinations are made on separate sub-samples. The data are in the file **reichpca.dat**.

Discussion

The first matter arising from the compositionally oriented analysis of the data concerns the correlations. The centred log-ratio values are listed in Table 4. We have used the log-ratio correlations for comparative reasons, but remind the reader about the reservations pertaining to the incoherency of product-moment correlations in simplex space. This array is to be compared with Table 2 in Reichart et al. (1997) – noting that the order of the entries in that table is not the same as that used in the rest of that paper.

TABLE 4
The Arabian Sea dust data. Centred log-ratio correlation coefficients
Centred log-ratio correlation matrix

	Al	Ba	Be	Ca	Fe	K	Mg	Mn	P	S	Sr	Ti	Y	Zn	Zr
1	1.000	−0.337	0.917	−0.422	0.510	0.974	0.609	0.762	−0.538	−0.636	−0.347	0.957	0.536	0.170	0.834
2	−0.337	1.000	−0.299	0.177	−0.190	−0.393	−0.657	−0.578	0.425	−0.152	0.043	−0.458	0.359	0.506	−0.397
3	0.917	−0.299	1.000	−0.424	0.572	0.878	0.561	0.697	−0.471	−0.580	−0.397	0.864	0.522	0.193	0.687
4	−0.422	0.177	−0.424	10.000	−0.717	−0.336	−0.184	−0.245	0.411	−0.285	0.937	−0.483	−0.042	−0.247	−0.251
5	0.510	−0.190	0.572	−0.717	1.000	0.443	0.245	0.308	−0.478	0.003	−0.675	0.501	0.192	0.298	0.246
6	0.974	−0.393	0.878	−0.336	0.443	1.000	0.699	0.813	−0.572	−0.659	−0.251	0.940	0.455	0.067	0.874
7	0.609	−0.657	0.561	−0.184	0.245	0.699	1.000	0.712	−0.470	−0.336	−0.078	0.652	−0.105	−0.293	0.645
8	0.762	−0.578	0.697	−0.245	0.308	0.813	0.712	1.000	−0.616	−0.445	−0.147	0.798	0.160	−0.245	0.710
9	−0.538	0.425	−0.471	0.411	−0.478	−0.572	−0.470	−0.616	1.000	0.002	0.379	−0.606	0.035	0.180	−0.510
10	−0.636	−0.152	−0.580	−0.285	0.003	−0.659	−0.336	−0.445	0.002	1.000	−0.299	−0.519	−0.695	−0.290	−0.598
11	−0.347	0.043	−0.397	0.937	−0.675	−0.251	−0.078	−0.147	0.379	−0.299	1.000	−0.402	−0.149	−0.314	−0.174
12	0.957	−0.458	0.864	−0.483	0.501	0.940	0.652	0.798	−0.606	−0.519	−0.402	1.000	0.440	0.111	0.861
13	0.536	0.359	0.522	−0.042	0.192	0.455	−0.105	0.160	0.035	−0.695	−0.149	0.440	1.000	0.553	0.381
14	0.170	0.506	0.193	−0.247	0.298	0.067	−0.293	−0.245	0.180	−0.290	−0.314	0.111	0.553	1.000	0.043
15	0.834	−0.397	0.687	−0.251	0.246	0.874	0.645	0.710	−0.510	−0.598	−0.174	0.861	0.381	0.043	1.000

Food for thought is provided by the results for the varimax rotated principal components, since these constitute the very basis for the scientific discussion of Reichart et al. (1997). Application of the program **pcaconst.exe** to the data yielded the results summarized in Table 5. These varimax rotated principal components are to be compared with the figures supplied in Table 3 of the Arabian Sea paper. The values for the inappropriate principal component analysis listed in Table 5 approximate closely those in Table 3 of Reichart et al. (1997). Comparison of the two sets of results in Table 5 for three factors disclose several major discrepancies to occur, these being of such magnitude as to place the factor-analytical conclusions of the published work in question.

TABLE 5
Comparison of varimax factor matrices using log-contrast principal components and the usual method.

	Log-contrast PCA factors			Usual PCA factors		
	1	2	3	1	2	3
Al	0.8360	0.2496	−0.4647	0.9573	0.1126	0.0495
Ba	−0.8158	−0.2654	0.4926	0.0688	0.9703	−0.0272
Be	−0.8134	−0.2441	0.4930	0.8912	0.1705	−0.0705
Ca	0.8266	0.2395	−0.4790	−0.7306	0.1536	0.1547
Fe	−0.8235	−0.2504	0.4669	0.6962	0.3007	−0.4884
K	−0.6887	−0.2215	0.6291	0.9625	0.0547	0.0694
Mg	0.8763	0.3353	−0.1928	0.6954	−0.2755	−0.1310
Mn	−0.8062	−0.2432	0.5062	0.8705	−0.1817	0.0099
P	0.7435	0.1914	−0.4023	−0.1062	0.3698	−0.2347
S	0.5150	0.3356	−0.7757	−0.0824	0.0377	−0.9807
Sr	−0.7884	−0.2836	0.4522	−0.5485	−0.0449	0.1497
Ti	0.8263	0.2482	−0.4591	0.9530	0.0403	−0.0460
Y	−0.8140	−0.2515	0.4987	0.6712	0.5293	0.0145
Zn	0.8368	0.2283	−0.4557	0.4969	0.5748	−0.2410
Zr	0.2618	0.9438	−0.1845	0.9064	0.0180	0.0742

Remarks

The analysis briefly reported here indicates that the evidence for "clusters of variables" reported by Reichart et al. (1997, p. 153), and provided with causal interpretations, is not free of blemishes for two reasons. Firstly, and most seriously, the constant-sum constrained has not been taken into account which can be expected to lead to pronounced distortions in the elements of the latent vectors, and this is indeed indicated by the differences in many of the elements obtained by the two sets of calculations. Secondly, the rôles of the parts in the interpretation of functional relationships depend on a reification of the rotated principal components, which is an exercise that cannot be undertaken without risk (reification of the principal component factor matrix is, as already noted, the mainspring of the published study). The examination of the centred log-ratio data matrix by cross-validation indicates several atypical and influential values to occur. This is also the case if the cross-validation exercise is carried out on the inappropriate matrix. The course of action open to the investigator in such a situation is to either delete the offending observations (in the interest of achieving stability in the latent vectors) or to use a method of robust estimation for principal components (cf. Reyment and Jöreskog, 1993). The plots of the relevant principal component scores clearly expose the presence of atypicalities in the data. The plots for the first two axes obtained by appropriate procedure is shown in Fig. 33. Inspection of this graph indicates that the points are widely spread in a manner such as to imply a considerable degree of heterogeneity in the sample.

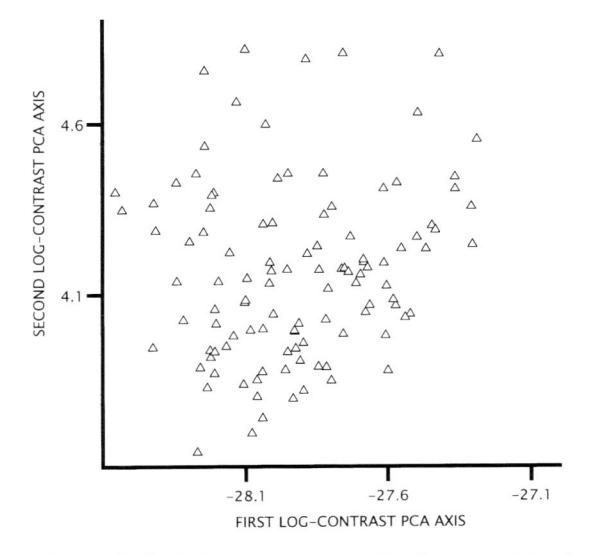

Fig. 33. Plot of the first and second principal component scores for the compositional principal component analysis of the dust data. It is obvious that there is much heterogeneity in the data.

SUGGESTED PROTOCOL FOR MULTIVARIATE ANALYSES OF GEOLOGICAL DATA

1 Inspect the array of observations by graphical means. *Graph Server* was constructed for such a purpose. Look for obviously atypical values since more marked deviations will in most cases distort results with respect to variances and, or, covariances. Check for heterogeneity in the dispersion pattern.

2 Are your data compositional in nature? Do the rows of the data-array have a constant sum; for example, frequencies sum to unity and percentages to 100. If so, you should use a method which is appropriate to simplex space. It is by no means uncommon to encounter data sets the multivariate analysis by the "usual procedure" of which gives about the same answer as the appropriate compositional method. However, there is no guarantee that this is always going to be the case and this variety of Russian Roulette is most emphatically not recommended.

3 If atypical observations are suspected (and in Geology this is rather the rule than the exception), then a cross-validatory analysis can then be undertaken in order to identify more closely atypical values and influential observations (that is, observations that are not obvious deviates, but which influence the stability of results. Jackknifing latent roots and vectors can be a very good means of diagnosing instability in estimates of canonical decompositions.

4 For most purposes, a simple *R*-mode principal component analysis will provide much valuable information. Alternatively, the inverted *Q*-mode analysis, such as is obtained by principal coordinates, can be tried. The pair-wise plots of the coordinate values are useful for disclosing discontinuities in the data. These methods are described in detail in Reyment and Jöreskog (1993), along with *R*–*Q*-mode procedures. Other multivariate analyses follow on naturally from this step.

5 If you believe you have trending data, be sure to test for a random walk (Bookstein, 1987) before drawing elaborate conclusions about evolutionary change in a fossil lineage, environmental tendencies manifested in borehole analyses and the like. For comparing complicated stratigraphical curves, cross-correlation is not a reliable technique for data which can not be located to accurately designated time-intervals. Gordon's (1973) method of slotting is preferable (see also Reyment, 1991).

SUMMARY

1 Evolution in *Echinocythereis*
   Data: echinreg.dat
   Program: multregr
   Data: echinpca.dat
   Program: pcomp2
2 Carbonates in CI chondrites
   Data: ivunapcd.dat, chondcva.dat, chonddfn.dat
   Programs: pcrdcons, cvaconst, dfnconst

3 Chemically induced variation in the Santonian ostracod *Veenia fawwarensis*
  Data: israelpc.dat
  Programs: covpc, cpca
4 Morphometrical relationships between scaphitid macroconchs and microconchs
  Data: discosc.dat, hoplo.dat
  Program: cva
5 A 225 km record of dust supply for the northern Arabian Sea.
  Data: reichpca.dat
  Program: pcaconst.
6 Suggested protocol for multivariate analyses of geological data

Glossary of computer program procedures

AITCHVAR: Computes the variation matrix for compositional data and the corresponding finite scale transformation of the variation matrix.

APPR: Computes approximate log-ratio means and variances from crude covariances and means.

BENZEC: Computes a simple correspondence analysis for a two-way contingency table (= a table of frequencies or proportions). This is a practical application of the singular value decomposition. Its main value lies with the exploitation of its graphical output.

CANCORR: Performs simple canonical correlation analysis and multivariate regression analyses. Useful for studying associations between sets of variables, such as occur in environmental work. The program provides simple canonical correlations as well as the Stewart–Love redundancy analysis, which is useful in cases where the data are not truly multivariate Gaussian and it is desired to reify the canonical coefficients.

CCRCONST: This program is the counterpart of *cancorr* for compositional data using the log-contrast canonical correlation.

COVMAT: Computes covariance and correlation matrices for 'open' and 'closed' data.

COVPC: Computes covariance matrices in a form suitable for input into the program for common principal components, *CPCA*.

CPCA: Common principal component analysis, a method designed for treating data with orientationally homogeneous covariance matrices (with respect to the axes of the ellipsoids of dispersion, but which may be differently inflated.

CPCCONST: Common principal component analysis for compositional data; the compositional counterpart of *CPCA*.

CVA: Carries out a canonical variate analysis encompassing generalized discriminant functions, a multivariate analysis of variance, generalized (Mahalanobis) statistical distances, tests of significance and the minimum spanning tree for canonical variate means.

CVACONST: The counterpart of *cva* for compositional data.

DFNCONST: Discriminant function analysis for compositional data. The counterpart of *disfun*.

DISFUN: Computes the linear and quadratic discriminant functions for two covariance matrices, the associated generalized distances and the Hotelling T^2 test of significance.

GABRIEL: The biplot for a rectangular matrix. A practical application of the singular value decomposition

HET: Computes the linear discriminant function and generalized distances for heterogeneous covariance matrices by the Anderson-Bahadur method and the Chernoff distance-decomposition criterion.

JKNFPCA: Computes jackknifed estimates of latent roots and vectors for covariance matrices or their associated correlation matrices.

LOGCOV: Computes the log-ratio covariance matrix.

MATINV: Illustrates the inversion of a square symmetric matrix.

MATOPS: Examples of some basic matrix operations

MULTEST: Routine for checking matrix multiplications with particular reference to the Biplot.

MULTNORM: A procedure for testing for multivariate skewness and kurtosis.

MULTREGR: Multiple regression analysis.

PCAIDENT: Demonstrates the usual (three) methods for scaling principal components.

PCACONST: Principal component analysis for compositional data. This is the counterpart of *pcomp2*.

PCOMP1: A simplified principal component analysis.

PCOMP2: A full principal component analysis with an option for principal component factor analysis and the Varimax rotation of the principal component factor axes.

PCOORD: Principal coordinate analysis with options for several kinds of association matrices, but favouring that of Gower (1971).

PCVALID: Program for principal component analysis using cross-validation techniques and jackknifed estimates of components and their standard errors and supplying an indication of the number of statistically interesting latent roots.

PCRDCONS: Constrained principal coordinate analysis, the counterpart of *pcoord* for compositional data.

PROPMAT: Program for testing a data-matrix for the constant-sum constraint.

SHRINKCV: Shrunken canonical variate analysis for achieving stability in canonical roots and vectors by Campbell's (1979) method.

SINGVAL: Computes a singular-value decomposition (= computes the Basic Structure) of a rectangular matrix, such as a data-matrix, thus providing, simultaneously, the principal axes of the matrix (= the *R*-mode solution) and the projections of the sample points onto these axes (the *Q*-mode solution). The *R*-mode solution corresponds to the columns of the rectangular matrix and the *Q*-mode solution to its rows.

SUBCOMP: Program for computing subcompositions for compositional data. An essential in any multivariate statistical analysis requiring a reduction in the number of parts to be treated.

N.B. The Compact Disk contains a file in Netscape Hypertext in which the input information for most of the programs is summarized, together with specimens of the input files.

References

Abe, K., Reyment, R. A., Bookstein, F. L., Honigstein, A., Almogi-Labin, A., Rosenfeld, A. and Hermelin, O., 1988. Microevolution in two species of ostracods from the Santonian (Cretaceous) of Israel. *Historical Geology*, 1, 303–322.

Aitchison, J., 1986. *The Statistical Analysis of Compositional Data.* Monographs on Statistics and Applied Probability. Chapman and Hall, London.

Aitchison, J., 1997. The one-hour-course in compositional data analysis or compositional data analysis is easy. *Proc. Math. Geol. Third Annual Conference, Barcelona* (September, 1997), (ed.), Vera Pawlowsky-Glahn, 3–35, Barcelona.

Anderberg, M. R., 1973. *Cluster Analysis for Applications*. Academic Press, London, xiii + 359 pp.

Anderson, T. W., 1963. Asymptotic theory for principal components analysis. *Ann. Math. Statist.*, 34, 122–148.

Anderson, T. W., 1984. *An Introduction to Multivariate Statistical Analysis.* 2nd edn, Wiley and Sons, New York.

Anderson, T. W. and Bahadur, R. R., 1962. Classification into two multivariate normal distributions with different covariance matrices. *Ann. Math. Statist.*, 33, 420–431.

Atkinson, A. C., 1985. *Plots, Transformations, and Regression.* Clarendon Press, Oxford.

Barceló, C., Pawlosky, V. and Grunsky, E., 1996. Some aspects of transformations of compositional data and the identification of outliers. *Math. Geol.*, 28, 501–518.

Barnard, M. M., 1935. The secular variation of skull characters in four series of Egyptian skulls. *Annals of Eugenics*, 6, 352–371.

Bellman, R., 1960. *Introduction to Matrix Analysis.* McGraw-Hill, New York.

Benzécri, J.-P., 1973. *L'Analyse des Données. 2. L'Analyses des Correspondances.* Dunod, Paris.

Blackith, R. E. and Reyment, R. A., 1971. *Multivariate Morphometrics*. Academic, London.

Bodergat, A. M, Carbonnel, G., Rio, M. and Keyser, D., 1993. Chemical composition of *Leptocythere psammophila* (Crustacea: Ostracoda) as influenced by winter metabolism and summer supplies. *Marine Biology*, 117, 53–62.

Bookstein, F. L., 1987. Random walk and the existence of evolutionary rates. *Paleobiology*, 13, 446–464.

Bookstein, F. L. and Reyment, T. A., 1992. Random walk and quantitative stratigraphical sequences. *Terra Nova*, 4, 147–151.

Bookstein, F. L., 1991. *Morphometric Tools for Landmark Data. Geometry and Biology.* Cambridge University Press, New York. xvii + 435 pp.

Borley, G. D., 1974. Oceanic islands. In *The Alkaline Rocks* (ed.), H. Sørensen, Wiley and Sons, New York, 311–330.

Cadrin, A. A. J., Kyser, T. K., Caldwell, W. G. E. and Longstaffe, F. J., 1996. Isotopic and chemical compositions of bentonites as paleoenvironmental indicators of the Cretaceous Western Interior Seaway. *Palaeogeog., Palaeoclim., Palaeoecol.*, 119, 301–320.

Campbell. N. A., 1979. Canonical Variate Analysis. Ph.D. Thesis, Imperial College, University of London.

Campbell, N. A., 1980. Shrunken estimators in discriminant and canonical variate analysis. *Applied Statistics*, 29,: 5–14.

Campbell, N. A., 1982. Robust procedures in multivariate analysis. II: robust canonical variate analysis. *Applied Statistics*, 31, 1–8.

Campbell, N. A., 1984. Canonical analysis with unequal covariance matrices: generalizations of the usual solution. *Math. Geol.*, 16, 109–124.

Campbell, N. A. and Reyment, R. A., 1978. Discriminant analysis of a Cretaceous foraminifer using shrunken estimators. *Math. Geol.*, 10, 347–359.

Campbell, N. A. and Reyment, R. A., 1980. Robust multivariate procedures applied to the interpretation of atypical individuals of a Cretaceous foraminifer. *Cretaceous Research*, 1, 207–221.

Chernoff, H., 1973. Some measures for discriminating between normal multivariate distributions with unequal covariance matrices. In P. R. Krishnaiah *Multivariate Analysis 3*, pp. 337–344. Academic Press, London.

Cook, R. D. and Weisberg, S., 1982. *Residuals and Influence in Regression.* Chapman and Hall, London.

Cooley, W. W. and Lohnes, P. R., 1971. *Multivariate Data Analysis.* Wiley and Sons, New York.

Cramér, H., 1946. *Mathematical Methods of Statistics.* Princeton University Press, N. J.

Davis, J. C., 1988. *Statistics and Data Analysis in Geology.* 2nd edn, Wiley and Sons, New York.

Davis, P. J., 1965. *The Mathematics of Matrices.* Blaisdell, New York.

Demange, J., Baubron, J.-C., Marcelot, G., Cotten, J. and, Maury, R. C., 1983. Cadre structural, pétrologie et géochimie de la série volcanique de Jabal al Abyad (Arabie Saoudite). *Bull. Centres Rech. Explor.-Prod. Elf-Aquitaine*, 7, 233–248.

Digby, P. G. and Kempton, R. A., 1987. *Multivariate Analysis of Ecological Communities.* Chapman and Hall, London.

Draper, N. R. and Smith, H., 1966. *Applied Regression Analysis.* Wiley and Sons Inc, N.Y., ix + 407 pp.

Eckart, C. and Young, G., 1936. The approximation of one matrix by another of lower rank. *Psychometrika*, 1, 211–218.

Efron, B., 1992. Jackknife and bootstrap standard errors and influence functions. *J. Roy. Statist. Soc.*, B, 54, 82–127.

Endo, K., Reyment, R. A. and Curry, G. B., 1995. Taxonomic relationships in *Terebratulina* (Brachiopoda) established by multivariate morphometrics. *Revista Esp. de Paleontología*, 10, 109–116.

Endress, M. and Bischoff, A., 1996. Carbonates in CI chondrites: clues to parent body evolution. *Geochemica et Cosmochimica Acta*, 60, 489–507.

Escoufier, Y., 1973. Le traitement des variables vectorielles. *Biometrics*, 29, 751–760.

Everitt, B., 1974. *Cluster Analysis.* Social Science Research Council, Heinemann, London.

Fisher, R. A., 1936. The use of multiple measurements in taxonomic problems. *Annals of Eugenics*, 7, 179–188.

Flury, B., 1984. Common principal components in *k* groups. *J. Amer. Statist. Assoc.*, 79, 892–898.

Flury, B., 1988. *Common Principal Components and Related Multivariate Models.* Wiley, New York.

Flury, B., 1995. Developments in principal component analysis. *Recent Advances in Descriptive Multivariate Analysis* (ed.), W. J. Krzanowski, 14–33.

Gabriel, K. R., 1971. The biplot display of matrices with application to principal components analysis. *Biometrika*, 58, 453–467.

Gabriel, K. R., 1995a. Biplot displays of multivariate categorical data. with comments on multiple correspondence analysis. *Recent Advances in Descriptive Multivariate Analysis* (ed.), W. J. Krzanowski, 190–226. Oxford Science Publications, Oxford.

Gabriel, K. R., 1995b. MANOVA biplots for two-contingency tables. *Ibid.*, 227–268.

Gantmcher, F. R., 1965. *Matrizenrechnung.* Vol. 1, xi + 324 pp; Vol. 2, viii + 244 pp. *VEB deutscher Verlag der Wissenschaften,* Berlin. Translation of *Teoriya Matrits* (1958).

Giresse, P., Maley, J. and Brenac, P., 1994. Late Quaternary palaeoenvironments in the Lake Barombi Mbo (West Cameroon) deduced from pollen and carbon isotopes of organic matter. *Palaeogeog, Palaeoclim., Palaeoecol.*, 107, 65–78.

Gleason, T. C., 1976. On redundancy in canonical analysis. *Psych. Bull.*, 83, 1004–1006.

Gnanadesikan, R., 1977. *Methods for Statistical Data Analysis of Multivariate Observations.* Wiley and Sons, New York.

Gordon, A. D., 1973. A sequence-comparison statistic and algorithm. *Biometrika*, 60, 197–200.

Gordon, A. D., 1981. *Classification.* Monographs on Applied Probability and Statistics. Chapman and Hall, London.

Gower, J. C., 1966. Some distance properties of latent root and vector methods used in multivariate methods. *Biometrika*, 55, 325–338.

Gower, J. C., 1971. A general coefficient of similarity and some of its properties. *Biometrics*, 27, 857–874.

Gower, J. C. and Hand. D. J., 1996. *Biplots.* Monographs on Statistics and Applied Probability 54, Chapman and Hall, London.

Greenacre, M. J., 1984. *Theory and Applications of Correspondence Analysis.* Academic Press, London, xi + 364 pp.

Hadi, A. S., 1992. Identifying multiple outliers in multivariate data. *J. Roy. Statist. Soc. B*, 54, 761–771.

Hampel, F. R., Ronchetti, E. W., Rousseeuw, P. J. and Statel, W. A., 1986. *Robust Statistics.* Wiley, New York, xxi + 502 pp.

Hand, D. J., 1981. *Discrimination and Classification.* Wiley and Sons, New York, x + 218 pp.

Harris, R. J., 1975. *A Primer of Multivariate Statistics.* Academic Press, London

Hill, M. O., 1974. Correspondence analysis: a neglected multi-variate method. *J. Roy. Statist. Soc. Ser. C*, 23, 340–354.

Hoaglin, D. C., Mosteller, F. and Tukey, J. W., 1985. *Exploring Data Tables, Trends, and Shapes.* Wiley, New York.

Hoorn, C., 1994. Fluvial palaeoenvironments in the intracratonic Amazonas Basin (early Mioceneearly Middle Miocene, Colombia). *Palaeogeog., Palaeoclim., Palaeoecol.*, 109, 1–50.

Hopkins, J. W., 1966. Some considerations in multivariate allometry. *Biometrics*, 22, 747–760.

Hotelling, H., 1936. Relations between two sets of variates *Biometrika*, 28, 321–377.

Jackson, J. E., 1991. *A User's Guide to Principal Components.* Wiley, New York, xxii+ 527 pp.

Jolliffe, I. T., 1986. *Principal Component Analysis.* Springer Verlag, New York.

Klingenberg, C. P., 1996. Multivariate allometry. *Advances in Morphometrics* (ed.), L. F. Marcus et al., 23–49.

Krzanowski, W. J., 1987a. Cross-validation in principal component analyses. *Biometrics,* 43, 575–584.

Krzanowski, W. J., 1987b. Selection of variables to preserve multivariate data structure, using principal components. *Applied Statistics,* 36, 22–33.

Krzanowski, W. J., 1988. *Principles of Multivariate Analysis.* Oxford Science Publications, Oxford.

Krzanowski, W. J., 1995. (ed.). *Recent Advances in Descriptive Multivariate Analysis.* Oxford Scientific Publications, Oxford.

Krzanowski, W. J. and Marriot, F. H. C., 1995. *Multivariate Analysis,* Part II, Arnold.

Lachenbruch, P. A. and Mickey, M. R., 1968. Estimation of error rates in discriminant analysis. *Technometrics,* 10, 1–11.

Laenen, B., Hertogen, J. and Vandenberghe, N., 1997. The variation of the trace-element content of fossil biogenic apatite though eustatic sea-level cycles. *Palaeogeog., Palaeoclim., Palaeoecol.,* 132, 325–347.

Lance, G. N. and Williams, W. T., 1965. Computer program for monothetic classification ('association analysis'). *Computing Journal,* 9, 246–249.

Landman, N. H. and Waage, K. M., 1993. Scaphitid ammonites of the Upper Cretaceous (Maastrichtian) Fox Hills Formation in South Dakota and Wyoming. *Bulletin of the American Museum of Natural History,* 215, 257 pp.

LeMaïtre, R. W., 1982. *Numerical Petrology.,* Elsevier, Amsterdam.

Mardia. K. V., 1970. Measures of multivariate skewness and kurtosis with applications. *Biometrika,* 57, 519–530.

Mardia, K. V., Kent, J. T. and Bibby, J. M., 1979. *Multivariate Analysis.* Academic Press, London, xv + 521 pp.

Mertens, B. J. A., 1998. Exact principal component influence measures applied to the analysis of spectroscopic data on rice. *Applied Statistics,* 47, 527–542.

Noll, P. D., Newsom, H. E., Leeman, W. P. and Ryan, J. G., 1996. The role of hydrothermal fluids in the production of subduction zone magmas: evidence from siderophile and chalcophile trace elements and boron. *Geochimica et Cosmochimica Acta,* 60, 587–611

Pearson, K., 1897. Mathematical contributions to the theory of evolution: on a form of spurious correlation which may arise when indices are used in the measurements of organs. *Proc. Roy Soc., London,* 60, 489–498.

Preisendorfer, R. W., 1988. *Principal Component Analysis in Meteorology and Oceanography.* Elsevier, Amsterdam.

Ramberg, J. S. and Schmeiser, B. W., 1972. An approximate method for generating symmetric random variables. *Comm. A. C. M.,* 15, 987–990.

Rao, C. R., 1949. On some problems arising out of discrimination with multiple characters. *Sankhya,* 9, 343–366.

Rao, C. R., 1952. *Advanced Statistical Methods in Biometrical Research.* Wiley and Sons, New York, xvii + 390 pp.

Rao, C. P. and Jayawardane, M. P. J., 1994. Major minerals, elemental and isotopic composition in modern temperate shelf carbonates, Eastern Tasmania, Australia: Implications for the occurrence of extensive ancient non-tropical carbonates. *Palaeogeog., Palaeoclim., Palaeoecol.*, 107, 49–63.

Reeside, J. B. and Cobban W. A., 1960. Studies of the Mowry Shale (Cretaceous) and contemporary formations in the United States and Canada. *Prof. Paper U.S. Geological Survey*, 355.

Reichart, G. J., den Dulk, M., Visser, H. J., van der Weijden, C. H. and Zachariasse, W. J., 1997. A 225 kyr record of dust supply, paleoproductivity and the oxygen minimum zone from the Murray Ridge (northern Arabian Sea), *Palaeogeog., Palaeoclim., Palaeoecol.*, 134, 149–169.

Reyment, R. A., 1969. A multivariate palaeontological growth problem. *Biometrics*, 22, 1–8.

Reyment, R. A., 1971. Multivariate normality in morphometric analysis. *Mathematical Geology*, 3, 357–368.

Reyment, R. A., 1972. Models for studying the occurrence of lead and zinc in a deltaic environment. In *Mathematical Models of Sedimentary Processes.* (ed.), T. W. Merriam, Plenum, New York.

Reyment, R. A., 1978a. Quantitative biostratigraphical analysis exemplified by Moroccan Cretaceous ostracods. *Micropaleontology*, 24, 24–43.

Reyment, R. A., 1978b. The interpretation of the smallest principal component. Vistelius Festschrift, *Spec. Publ. Akademia Nauk USSR*, 163–168. (In Russian.)

Reyment, R. A., 1985. Phenotypic evolution in a lineage of the Eocene ostracod *Echinocythereis*. *Paleobiology*, 11, 174–194.

Reyment, R. A., 1991. *Multidimensional Palaeobiology*, Pergamon Press, Oxford, ix + 377 pp.

Reyment, R. A., 1996. Case study of the statistical analysis of chemical compositions exemplified by ostracod shells. *Environmetrics*, 7, 39–47.

Reyment, R. A., 1997. Multiple group principal component analysis. *J. Math. Geol.*, 29, 197–224.

Reyment, R. A. and Brännström, B., 1962. Certain aspects of the physiology of *Cypridopsis* (Ostracoda, Crustacea). *Stockh. Contr. Geol.*, 9, 207–242.

Reyment, R. A. and Jöreskog, K. G., 1993. *Applied Factor Analysis in the Natural Sciences.* Cambridge University Press, New York. Second printing (1996).

Reyment, R. A. and Kennedy, W. J., 1998. Taxonomic recognition of species of *Neogastroplites* (Ammonoidea, Cenomanian) by geometric morphometric methods.*Cretaceous Research* 18, 1–18.

Rohlf, F. J. and Sokal, R. R., 1969. *Statistical Tables.* Freeman, San Francisco.

Schoenberg, I. J., 1935. Remarks to Maurice Fréchet's article "Sur la définition axiomatique d'une classe d'espaces distanciés vectoriellement applicable sur l'espace de Hilbert". *Ann. Math (Second Series)*, 36, 724–732.

Schönemann, P. H. and Carroll, R. M., 1970. Fitting one matrix to another under choice of a central dilation and a rigid motion. *Psychometrika*, 35, 245–256.

Searle, S. R., 1966. *Matrix Algebra for the Biological Sciences.* Wiley, New York.

Seber, G. A. F., 1984. *Multivariate Observations*, Wiley, New York.

Sigurdsson, H., Bonté, P., Turpin, L., Chaussidon, M., Metrich, N., Pradel, P. and D'Hondt, S., 1991. Geochemical constraints on source region of Cretaceous/Tertiary impact glasses. *Nature*, 353, 839–842.

Sørensen, H., 1974. *The Alkaline Rocks.* Wiley and Sons, New York.

Stewart, D. K. and Love, W. A., 1968. A general canonical correlation index. *Psychological Bulletin*, 70, 160–163.
Telnaes, N., Björseth, A., Christy, A. A., and Kvalheim, O. M., 1987. Interpretation of multivariate data: Relationship between phenanthrenes in crude oils. *Chemometrics and Intelligent Laboratory Systems*, 2, 149–153.
ter Braak, C. F. J., 1987. The analysis of vegetation-environment relationships by canonical correspondence analysis. *Vegetation*, 69, 69–77.
Thiede, J., Nees, S., Schulz, H. and De Deckker, P., 1997. Oceanic surface conditions on the sea floor of the Southwest Pacific Ocean through the distribution of foraminifers and biogenic silica. *Palaeogeog., Palaeoclim., Palaeoecol.*, 131, 207–239.
Tribollivard, N. P., Desprairies, A., Lallier-Vergès, E., Bertrand, P., Moureau, N., Ramdoni, A. and Ramanampisoa, L., 1994. Geochemical study of organic-rich matter from the Kimmeridge Clay Formation of Yorkshire (U.K.): productivity versus anoxia. *Palaeogeog., Palaeoclim., Palaeoecol*, 108, 165–181.
Usui, A., 1992. Hydrothermal manganese minerals in Leg 126 cores. *Proc. Ocean Drilling Program, Scientific Results*, 126, 113–123.
Whitten, E. H. Y., 1993. A solution to the percentage data problem in petrology. In *Computers in Geology,* (ed.), J. C. Davis and U. C. Herzfeld, *Studies in Mathematical Geology*, No. 5, Oxford University Press, Oxford, pp. 195–206.
Whitten, E. H. T., 1995. Open and closed compositional data in petrography. *Mathematical Geology,* 27, 789–806.
Zurmühl, R., 1964. Matrizen und ihre technischen Anwendungen. Springer-Verlag, Berlin, Göttingen, Heidelberg. 452 pp.

Index

This index is designed to be used in conjunction with the indices of methods, programs and examples provided at the end of each methodological chapter. Such information has not been repeated in the following and must be sought in the appropriate locations:

Abe, K. 248, 275
Afrobolivina afra foraminiferal data 114, 116, 137, 206
Aitchison, J. 14, 17, 18, 20, 21–26, 28, 30, 32–35, 131, 140, 168, 169, 182, 187, 200, 218, 225, 233, 234, 246, 275
Alkaline rocks, Atlantic and Pacific Oceans 183, 185
Allometric model of common principal component analysis 197, 203
Almogi-Labin, A. 275
Amalgamation of compositional data 19
Anderberg, M. R. 137, 275
Anderson, T. W. 4, 155, 166, 193, 211, 220, 275
Anderson-Bahadur T^2 163, 166, 167
ANOVA, of shape-indicators 179, 184
Approximation for compositional arrays 34
Arabian Sea dust example 264
Association matrices 11, 134
Atkinson, A. C. 224
Atlas Mountains Cretaceous 206
Atypical observations 128, 204, 267
Autonne, L. 141

Bahadur, R. R. 163
Baltic Sea ostracods 169, 185
Barceló, C. 15, 35, 275
Barnard, M. M. 155, 275
Bases of compositional data 18
Basic structure of a matrix 141
Baubron, J. C. 276
Bellman 104, 191
Bentonite data (Cretaceous, USA) 220, 231
Benzécri, J-P. 9, 145, 275
Bertrand, A. 280
Bibby, J. M. 4, 278
Biplots 9, 142, 148, 149

Bischoff, A. 36, 244, 245, 247, 248, 276
bitmap 54
Björseth, A. 280
Blackith, R. E. 174, 275
Bodergat, A. M. 169, 185, 188, 275
Bonté, P. 279
Bookstein, F. L. 7, 223, 268, 275
Borley, G. D. 133, 275
Box-Cox transformation 35
Brachiopod data (*Terebratella*) 161, 164
Brännström, B, 195, 198, 279
Brenac, P. 277

Cadrin, A. A. J. 220, 231, 232, 238, 276
Cameroun 5
Campbell, N. A. 35, 174, 175, 182, 204, 206, 209, 213, 272, 276
Caldwell, W. G. E. 276
Canada 232
Canary Islands volcanics data 140
Canonical correlation 9, 211, 226
  log-contrast 232, 234
CANOCO 9
Canonical variate analysis 9, 174 255
  Stability of estimates of 204
  Of compositions 182, 245
Carbonnel, G. 185, 275
Carroll, R. M. 124, 279
Centred log-ratio covariance matrix 24, 131
Chaussidon, M. 279
Chayes Kruskal method 21
Chemometrics 121
Chernoff, H. 276
Chernoff's distance criterion 167
Chondrites 36, 244
Cobban, W. A. 176, 279
Coherence of subcompositions 21

Common principal component analysis 190, 248, 249
  Adequacy of model 192
  Compositional 199
  Standard errors 193
Components, parts recorded as 6
Compositional data, definition 14, 15
Computing requirements for programs 2
Constant sum condition 30
Constant weight stratagem 17
Cook, R. D. 224, 276
Cooley, W. W. 212, 218, 276
Correlation matrix 11
Correlation and compositions 32
Correlation, spurious 5
Correspondence analysis 9, 142, 145
Christy, A. A. 280
Cramér, H. 34, 276
Cotton, J. 276
Covariance matrix 11
Cross validation 120
  Constrained 123
Curry, G. B. 276
Cuxhaven 185
Cypridopsis (ostracod) data 195

Data categories 1
Data matrix defined 8, 10
Davis, J. C. 3, 276
Davis, P. J. 4, 276
De Deckker, P. 280
Demange, J. 225–228, 276
Desprairies, A. 180
D'hondt, S. 279
Digby, P. G. 134, 276
Discoscaphites (ammonite) 254
Discriminant functions 9, 168, 169, 259
DLL 43
Dolomite in chondrites 245
Draper, N. R. 222, 276
Dulk, M. den 279
Dust, wind transported 265

Echinocythereis (ostracod) data 157, 214, 239
Eckart-Young theorem 134, 141
Eckart, C. 276
Efron, B. 35, 121, 276
emf 53
Endo, K. 162, 164, 276
Endress, M. 36, 244, 276
Environmental chemistry 185, 195

Environmetrics 150, 186
Epicontinental flooding (Cenomanian-Turonian) 232
Escoufier, Y. 213, 276
Everitt, B. 276
Evolution in Eocene ostracods 239

Factor analysis 104, 113
Finite scale transformation of variation matrix 30
Fisher, R. A. 9, 155, 159, 276
Fixed mode multivariate analysis 103, 113
Flury, B. 190, 193, 194, 199, 277

Gabriel, K. R. 10, 142, 145, 148, 150, 277
Galton, F. 11, 16
Gantmacher, F. R. 4, 104, 277
Generalized statistical (Mahalanobis) distance 9, 188
Giresse, P. 5, 277
Gleason, T. C. 218, 277
Gosset, W. S. 8
Gnanadesikan, R. 35, 277
Gordon, A. D. 11, 14, 35, 135, 149, 150, 268, 277
Gower, J. C. 8, 11, 133, 135, 139, 246, 272, 277
Graph Server 40
  simple example 153
Graph Wizard 96
  example of use 153
Greenacre, M. J. 145, 277
Grunsky, E. 275
GS language 56
GS tutorial 75

Hadi, A. S. 35, 277
Haitian bolide data 132, 133
Hampel, F. R. 35, 277
Hand, D. J. 155, 167, 277
Harker diagram 18
Harris, P. J. 4, 211, 264, 277
Hertogen, J. 278
Heterogeneous generalized statistical distance 163, 168
Hierarchy of statistical methods 3
Hill, M. O. 145, 277
Hoaglin, D. C. 175, 277
Honigstein, A. 275
Hoorn, C. 5, 277
Hopkins, J. W. 197, 203, 277

Index

Hoploscaphites (ammonite) 254, 259
Hotelling, H. 213, 277
Hotelling's T^2 9, 156, 159
HTML documentation 39

Influential observations 128, 204
Invariant principal component 234, 260

Jabal al Abyad (Saudi Arabian rift) 225
Jackknifing technique 35, 121 129, 243
Jackson, J. E. 103, 104, 111, 113, 150, 219, 278
Jayawardane, M. P. J. 6, 279
Japan, 31, 129
Jolliffe, I. T. 103, 150, 278
Joreskog, K. G. 4, 5, 12, 17, 103, 104, 106, 113, 120, 122, 133, 141, 143, 185, 243, 267, 268, 279

Keijella ostracod data 129
Kempton, R. A. 134, 276
Kennedy, W. J. 176, 182, 279
Keyser, D. 185, 275
Kimmeridge Clay 6
Klingenberg, C. P. 195, 197, 203, 278
Kvalheim, O. 289
Krzanowski, W. J. 35, 121, 156, 169, 204, 209, 213, 234, 278

Lachenbruch, P. A. 121, 278
Laenen, B. 10, 278
Lallier-Vergès, A. 280
Lance, G. N. 137, 278
Landman, N. H. 254, 264, 278
Leeman, W. P. 278
LeMaitre, R. W. 169, 278
Leptocythere (ostracods) data 150, 169, 185
Linear discriminant function 155
Logarithmic growth spiral 256
Log-contrast canonical components 219
Log-contrast canonical variate analysis 187
  Discriminant function 247
  Of a D-part composition 131
Log-ratio covariance matrix 24
Log-ratio variance 22
Lohnes, P. R. 212, 218, 276
Longstaffe, F. J. 276
Love, W. A. 218, 280

Macroconch ammonites 254
Magnesium and ecophenotypy in fossil ostracods 253

Mahanalobis, P. C. 9, 149, 168
Major product moment 141
Maley, J. 277
MANOVA 9, 185
Marcelot, G. 276
Mardia, K. V. 36, 278
Marriot, F. H. 156, 169, 278
Maury, R. C. 276
Mertens, B. J. A. 172, 278
metafile 52
Metrich, M. 279
Mickey, M. R. 121, 278
Microconch ammonites 254
Minor product moment 141
Minimum spanning tree 136, 175, 227, 258, 263
Monsoonal circulation study 264
Moore-Penrose inverse 111, 169
Morocco (Cretaceous) 206, 224
Morphometrics 177
Mosteller, F. 277
Moureau, P. 289
Multiple discriminant function 188
Multiple regression example 239
Multivariate normality 36

Nees, S. 280
Negative bias of correlations 20
Neogastroplites (ammonite) data (USA) 176, 223
Newsom, H. E. 278
Nigeria 116, 137
Noll. P. P. 16, 278
North Sea oils 146, 147
Null correlation problem 21

Open set confusion 20
Ordinated scores 10, 189
OTA (acronym) 258
Outliers 35, 178

Palaeo-oceanography, example of 232
Parts (as opposed to variables) 12, 15
Pawlowsky, V. 275
Pearson, K. 5, 8, 11, 16, 213, 226, 278
Pearson's coefficient of racial likeness 7
Phenanthrenes 146–148
Population, statistical 7
Pradel, P. 279
Preisendorfer, R. W. 103, 113, 278

PRESS, acronym 127, 128
Principal component analysis 8, 104, 144, 241, 255
    factor analysis 113, 119, 265
    log-contrast 232
Principal coordinate analysis 8, 133, 144
principal coordinate analysis, compositional 139, 226, 230, 245
Prim, R. C. 175
Procrustean fitting 124
Proportionality between parts 33

Quadratic discriminant function 156, 161, 247
Q-mode methods 133
Q–Q probability plots 175, 177, 245

Ramanampisoa, A. C. 280
Ramdoni, A. 280
Random mode methods 103, 113
Random walk 7
Rao, C. P. 6, 279
Rao, C. R. 156, 278
Ratio plots 226
Redundancy analysis 213, 217, 218, 237
Reeside, J. B. 176, 279
Reichart, G. J. 264, 266, 267, 279
Reification 13, 173, 180, 182, 212, 217, 267
Relative variance 32
Replicates 16
Reyment, R. A. 4, 5, 7, 12, 17, 35, 103, 104, 106, 113, 120 122, 133, 141, 143, 150, 157, 166, 169, 174, 176, 182, 185, 186, 191, 195, 197–199, 204, 206, 209, 219, 239, 243, 267, 268, 275, 276, 279
Rio, G. 135, 275
Robust principal components 267
R-mode methods 104
Rose diagram 5
Rosenfeld, A. 275
Rotation of latent vectors 114, 116
Rohlf, F. J. 37, 279
Ronchetti, E. W. 277
Roscoff 169, 185
Rousseeuw, P. J. 277
Ryan, J. G. 278

Saudi-Arabian rift volcanics 225, 226, 229, 230
Scaling latent vectors, methods of 108
Scaphitids (ammonites) 254, 264
Schmeiser, B. W. 175, 278
Schoenberg, P. 135, 279
Schöneman, P. H. 124, 279

Shrunken canonical estimators 205, 206, 208
Schulz, H. 280
Searle, S. R. 4, 36, 279
Seber, G. A. F. 4, 161, 193, 204, 209, 279
SendFile 47
Shiloah (Israel), Santonian 248
Sigurdsson, H. 12, 132, 279
Simplex 17, 20
Singular value decomposition 8, 141, 142
Smith, H. 222, 276
Sokal, R. R. 37, 279
Srrensen, H. 122, 133, 140, 183, 226, 279
South Dakota (USA) Cretaceous 254
Spain (Aragonese Eocene) 157, 214
Stable estimates of latent vectors 129, 174
Statel, W. A. 277
Stewart, P. K. 218, 280
Subcompositions 19, 29
Sylvester, J. J. 141

Telnaes, N. 146, 280
ter Braak, C. J. F. 9, 280
Thiede, J. 134, 280
Tokyo Bay, ostracod study 129
Tribollivard, N. P. 6, 280
Turpin, L. 279
Tukey, J. W. 121, 277

Usui, A. 31, 280

Vandenberghe, N. 278
Variables
    Binary 14
    Continuous 14
    Discrete 14
    Nominal 14
    Ordinal 14
    Qualitative 14
Variation matrix 24, 29
Vector correlation 213
Veenia (ostracods) data 206, 224
Veenia (ostracods, Israel) 248
Visser, H. J. 279

Waage, K. M. 254, 264, 278
Weierstrass, K. T. W. 212
Weijden, C. H., van der 279
Weisberg, S. 224, 276
Western Interior Basin, USA (Cretaceous) 232
Whitten, E. H. T. 17, 280

Index

Williams, W. T. 137, 278
wmf 53

Yorkshire Jurassic 6
Young, G. 141, 276

Zachariasse, W. J. 279
Zero data entries 22
Zero latent roots in principal components 120
z-scores and y-scores 112
Zurmühl, R. 4, 280